Haynes Guide
The Complete Handbook

Yellowstone National Park

733

Haynes-Photo.

GIANT GEYSER—THE HIGHEST GEYSER IN THE WORLD—250 FT.

HAYNES GUIDE

THE COMPLETE HANDBOOK

YELLOWSTONE NATIONAL PARK

Descriptive - Geological - Historical

Compiled by
JACK E. HAYNES, B. A.

Thirtieth Annual Edition
REVISED AND ENLARGED

1916

Published and Illustrated by

Haynes
ST PAUL

Official Photographer of Yellowstone National Park

ST. PAUL, MINN.

COPYRIGHTED 1912
COPYRIGHTED 1913
COPYRIGHTED 1914
COPYRIGHTED 1915
BY F. JAY HAYNES
COPYRIGHTED 1916
BY J. E. HAYNES

The Pioneer Company
Printers
Saint Paul

Preface

THE purpose of this book is to guide the tourist on his tour of Yellowstone National Park, and to make his visit pleasant and interesting. To this end it names, describes, and pictures all the points of interest in the park and presents in concise and readable form the scientific and historical information necessary to a clear understanding of the various phenomena.

The trip through the park, roughly speaking, is in the form of a circle, and the descriptions, illustrations, and maps in this book are arranged in such a manner as to form a perfect guide from each and every one of the several entrances.

To know how to use this book you need only turn to the place indicated for your entrance (see page 23).

CONTENTS

PANORAMA OF YELLOWSTONE NATIONAL PARK.

YELLOWSTONE NATIONAL PARK.

The Yellowstone Park is the first established and largest of all our National Parks. It was dedicated March 1, 1872, by an Act of Congress, as a park "For the Benefit and Enjoyment of the People."

In this park, within an ellipse fifty miles long by forty miles wide, the traveler may see such a variety of natural wonders as he can otherwise see only by circling the globe. The geysers at the Upper Basin are greater in number and in size than the famous geysers in Iceland. The Grand Canyon of the Yellowstone, with its two waterfalls aggregating over 400 feet and its wonderful coloring, are not surpassed in beauty by the Grand Canyon of the Colorado or by Niagara Falls. Yellowstone Lake, with the exception of Lake Titicaca in the Peruvian Andes, is the largest lake in the world at its altitude of 7,741 feet. The great terraces at Mammoth Hot Springs, both in size and in beauty of coloring, surpass those in New Zealand.

Among the many other features of interest are the Obsidian Cliff, an escarpment of volcanic glass; the petrified trees; the grotesque forms of the volcanic rock; the varicolored hot and cold pools; the Paint Pots, Mud Volcano, and the caves exhaling poisonous gas which destroys both bird and animal life; the wild animals in their natural habitat, and the streams abounding in native trout.

The excellent highways that connect the points of interest were built by the United States Engineer Corps solely for the pleasure and convenience of tourists. In their construction no effort or expense has been spared to penetrate the natural barriers that formerly made many of these wonders inaccessible.

The Park is under the jurisdiction of the Department of the Interior—the Commanding Officer at Fort Yellowstone is Acting Superintendent.

United States cavalrymen preserve order, patrol the roads, and guard the objects of interest from vandalism. Government scouts protect the game from poachers. When the winters are severe the animals are fed and every means is taken to keep them from suffering.

How to Reach the Park.—Two railroads have built branch lines to the park boundaries: The Union Pacific System to Yellowstone, Montana, at the Western Entrance, and the Northern Pacific Railway to the Northern Entrance at Gardiner, Montana. The Chicago, Burlington & Quincy Railroad has a terminus at Cody, Wyoming, from which a highway leads to the Eastern Entrance. Another highway from Jackson's Hole, south of the park, gives access to the Southern Entrance.

How to Make the Park Trip.—On August 1, 1915, private automobiles were permitted, under proper regulations, to make the complete tour of the park. In 1916 the first motor company was organized to carry passengers from Cody, Wyoming, to the Yellowstone Lake and return. Transportation in the park proper during 1916 is by four and six-horse coaches, two-horse surreys, and saddle and pack outfits.

The system of seven hotels is operated throughout the park by the Yellowstone Park Hotel Company under lease from the Department of the Interior during the annual tourist season, June 15th to September 15th.

The Yellowstone-Western Stage Company, and the Yellowstone Park Transportation Company operate in connection with the hotels, giving daily service from the Western Entrance at Yellowstone, Montana, and the Northern Entrance at Gardiner, Montana, respectively.

The two largest camping companies are the Wylie Permanent Camping Company and the Shaw & Powell Camping Company. They transport passengers from the various entrances and furnish both noon and night accommodations in permanent camps.

Hotels, camps and stores are all owned by private individuals or companies. Concessions are granted by the Department of the Interior, which regulates all rates and charges in the park.

A PARK ANTELOPE.

1—578 r.

DEPARTMENT OF THE INTERIOR.
FRANKLIN K. LANE, SECRETARY.

REGULATIONS GOVERNING THE ADMISSION OF AUTOMOBILES INTO THE YELLOWSTONE NATIONAL PARK FOR THE SEASON OF 1916.
(EFFECTIVE JUNE 15, 1916.)

WASHINGTON, D. C., *March 1, 1916.*

Pursuant to authority conferred by section 2475, Revised Statutes, United States, and the act of Congress approved May 7, 1894, the following regulations governing the admission of automobiles into the Yellowstone National Park are hereby established and made public:

1. **Automobiles.**—The park is open only to such automobiles as are operated for pleasure and not to those carrying passengers who are paying, either directly or indirectly, for the use of the machine.

2. **Motorcycles.**—Motorcycles are not permitted to enter the park.

3. **Tickets of Passage.**—Ticket of passage must be secured and paid for at the checking station where the automobile enters the park. This ticket must be conveniently kept, so that it can be exhibited to park guards on demand, and must be surrendered at the last checking station on leaving the park. Tickets of passage will show (a) name of owner, (b) license number of automobile, (c) name of State issuing license, (d) make of machine and manufacturer's number, (e) name of driver, (f) seating capacity of machine, and (g) number of passengers.

4. **Fees.**—Fees are payable in cash only, and will be as follows: $7.50 for a single trip through the park, and $10 for the season. All permits will expire on October 1 of the year of issue.

5. **Muffler cut-outs.**—Muffler cut-outs must be closed while approaching or passing riding horses, horse-drawn vehicles, hotels, camps, or soldier stations.

6. **Distance Apart—Gears and Brakes.**—Automobiles while in motion must not be less than 50 yards apart, except for purpose of passing, which is only permissible on comparatively level or slight grades. All automobiles, except while shifting gears, must retain their gears constantly enmeshed. Persons desiring to enter the park in an automobile will be required to satisfy the guard issuing the ticket of passage that the machine in general, and particularly the brakes and tires, are in first-class working order and capable of making the trip, and that there is sufficient gasoline in the tank to reach the next place where it may be obtained, and carry two extra tires. For this purpose, all drivers will be required effectually to block and skid the rear wheels with either foot or hand brake, or such other brakes as may be a part of the equipment of the automobile. Gasoline can be purchased at regular supply stations as per posted notices.

7. **Speeds.**—Speeds must be limited to 12 miles per hour ascending and 10 miles per hour descending steep grades, and to 8 miles per hour when approaching sharp curves. On good roads with straight stretches, and when no team is nearer than 200 yards, the speed may be increased to 20 miles per hour. Horns must be sounded at all curves where the road can not be seen for at least 200 yards ahead, and when approaching teams or riding animals.

8. **Teams.**—When teams, saddle horses, or pack trains approach, automobiles will take the outer edge of the roadway, regardless of the direction in which they may be going, taking care that sufficient room is left on the inside for the passage of vehicles and animals. Teams have the right of way, and automobiles will be backed or otherwise handled as may be necessary so as to enable teams to pass with safety. In no case must automobiles pass animals on the road at a greater speed than 8 miles per hour.

9. **Fines.**—Fines or other penalties will be imposed for arrival of automobiles at any point before approved lapse of time, hereinafter given, at the following rates: $0.50 per minute for each of first five minutes; $1.00 per minute for each of the next 20 minutes; $25.00 fine or ejection from the park, or both, in the discretion of the Acting Superintendent of the park, for being more than 25 minutes early

AUTOMOBILE REGULATIONS—Continued.

10. **Penalties.**—Violation of any of the foregoing rules or general regulations for government of the park will cause revocation of ticket of passage, and in addition to the penalties hereinbefore indicated will subject the owner of the automobile to any damage occasioned thereby, immediate ejectment from the reservation, and be cause for refusal to issue new ticket of passage to the owner without prior sanction in writing from the Secretary of the Interior.

11. **Accidents.**—When, due to breakdowns or accidents of any other nature, automobiles are unable to keep going or to reach the next stopping place on time, they must be immediately parked off the road, or where this is impossible, on the outer edge of the road, and wait until the next schedule for automobiles past that point, or until given special permission to proceed by park guards.

12. These regulations and schedules do not apply to automobiles passing over the county road in the northwest corner of the park, en route to the town of Yellowstone, Montana.

Approved: R. B. Marshall,

    Stephen T. Mather,   *Superintendent of National Parks.*

    Assistant to the Secretary.

SCHEDULES AND GENERAL INSTRUCTIONS.

Automobiles may leave the park by any one of the authorized routes of entrance. Automobile drivers should compare their watches with the clocks at checking stations.

Automobiles stopping over at points other than the hotels and permanent camps will be allowed to resume travel only at such times as permits them to fall in with a subsequent regular automobile schedule past the point of stop-over. Such automobiles while stopping over must park out of sight of, or at least 100 yards from, the main road.

Automobiles stopping over at permanent camps must leave the same at the proper time to conform with the published schedules from the nearest hotels. Detailed times of departure to comply with this provision will be posted at the particular camps concerned.

When, due to breakdowns or accidents of any other nature, automobiles are unable to keep going, or to reach the next stopping place on time, they must be immediately parked off the road, or where this is impossible, on the outer edge of the road, and wait until the next schedule for automobiles past that point, or until given special permission to proceed by park guards.

Automobiles will not be permitted for use on local trips around hot springs formations or other points of interest off the main roads, except in the case specially noted at Artist Point, in the morning schedule from the Lake Hotel to Canyon.

Speeds.—Speeds must be limited to 12 miles per hour ascending and 10 miles per hour descending steep grades, and to 8 miles per hour when approaching sharp curves. On good roads with straight stretches, and when no team is nearer than 200 yards, the speed may be increased to 20 miles per hour.

Horn.—The horn will be sounded on approaching curves or stretches of road concealed for any considerable distance by slopes, overhanging trees, or other obstacles; and before meeting or passing other machines, or riding or driving animals.

Teams.—When teams, saddle horses, or pack trains approach, automobiles will take the outer edge of the roadway; regardless of the direction in which they may be going, taking care that sufficient room is left on the inside for the passage of vehicles and animals. Teams have the right of way, and automobiles will be backed or otherwise handled as may be necessary so as to enable teams to pass with safety. In no case will automobiles pass animals on the road at a greater speed than 8 miles per hour.

In addition to the schedules herein given, automobiles must keep clear of any horse-drawn passenger vehicles running upon regular schedules which may be following them; upon overtaking any horse-drawn passenger vehicles running upon regular schedules, automobiles must not attempt to pass or approach closer than within 150 yards of the same.

Reduced engine power—Gasoline, etc.—Due to the high altitude of the park roads, averaging nearly 7,650 feet for the belt line and east, north, and west entrances, the power of all automobiles is much reduced, so that about 50 per cent more gasoline will be required than for the same distance at lower altitudes. Likewise one lower gear will generally have to be used on grades than would have to be used in other places. A further effect that must be watched is the heating of the engine on long roads, which may become serious unless care is used. Gasoline can be purchased at regular supply stations as per posted notices.

AUTOMOBILE SCHEDULE

	Miles.	Schedule A.		Schedule B.	
		Not earlier than—	Not later than—	Not earlier than—	Not later than—
GARDINER TO NORRIS.					
Leave Gardiner Entrance	0	6.00 a. m.	6.30 a. m.	2.30 p. m.	3.00 p. m.
Arrive Mammoth Hot Springs	5	6.30 a. m.	7.00 a. m.	2.50 p. m.	3.30 p. m.
Leave Mammoth Hot Springs	0	6.45 a. m.	7.15 a. m.	5.45 p. m.	6.15 p. m.
Leave 8-mile Post	8		8.00 a. m.	See Note 3.	
Arrive Norris	20	8.30 a. m	9.00 a. m.		
NORRIS TO WEST ENTRANCE.					
Leave Norris	0			4.00 p. m.	4.30 p. m.
Arrive West Entrance	27			6.00 p. m.	6.30 p. m.
NORRIS TO CANYON.					
Leave Norris	0			2.15 p. m.	2.30 p. m.
Arrive Canyon	11			3.00 p. m.	3.30 p. m.
(For Gallatin Station Entrance see Note 1.)					
NORRIS TO FOUNTAIN HOTEL.					
Leave Norris	0	8.30 a. m.	9.15 a. m.	4.00 p. m.	4.30 p. m.
		(Via Mesa Road.)		(Via Mesa Road or Madison Junction.)	
Leave Firehole Cascades	14.7		10.30 a. m.		
Arrive Fountain Hotel	20	10.30 a. m.	11.00 a. m.	5.45 p. m.	6.15 p. m.
(For Gallatin Station Entrance see Note 1.)					
WEST ENTRANCE TO FOUNTAIN HOTEL.					
Leave West Entrance	0	6.45 a. m.	7.15 a. m.	7.30 p. m.	8.00 p. m.
Arrive Fountain Hotel	21	8.30 a. m	9.00 a. m.	See Note 3.	
FOUNTAIN HOTEL TO THUMB.					
Leave Fountain Hotel	0	10.30 a. m.	11.00 a. m.	5.45 p. m.	6.15 p. m.
Arrive Upper Basin (Old Faithful Inn)	9	12.00 m.	12.30 p. m.	6.45 p. m.	7.00 p. m.
Leave Upper Basin (Old Faithful Inn)	0	2.30 p. m.	3.00 p. m.	7.00 a. m.	7.30 a. m.
Arrive Thumb Station	19	4.30 p. m.	5.00 p. m.	8.30 a. m.	9.30 a. m.
(For South Entrance see Note 1.)					
THUMB TO LAKE HOTEL.					
Leave Thumb Station	0	4.30 p. m.	5.00 p. m.	8.30 a. m.	9.30 a. m.
Arrive Lake Hotel	15	5.45 p. m	6.15 p. m.	10.00 a. m	11.30 a. m.
LAKE HOTEL TO EAST BOUNDARY.					
Leave Lake Hotel (see Note 1)	0				
Arrive East Boundary	28				
EAST BOUNDARY TO LAKE HOTEL.					
Leave East Boundary (see Note 1)	0				
Arrive Lake Hotel	28				

AUTOMOBILE SCHEDULE— Continued.

	Miles	Schedule A.		Schedule B.	
		Not earlier than—	Not later than—	Not earlier than—	Not later than—
LAKE HOTEL TO CANYON.					
Leave Lake Hotel	0	7.00 a. m.	7.30 a. m.	2.00 p. m.	2.30 p. m.
(See Note 2.)					
Leave Canyon Soldier Station	16	9.00 a. m.	10.00 a. m.		
Arrive Canyon Hotel	17	9.10 a. m.	10.10 a. m.	3.15 p m.	3.45 p. m.
CANYON TO NORRIS.					
Leave Canyon Hotel	0	2.15 p. m.	2.30 p. m.		
Arrive Norris	12	3.00 p. m.	3.30 p. m		
(For Schedules from Norris to Fountain, Upper Basin, and West Entrance, see page 8.)					
CANYON TO TOWER FALLS.					
Leave Canyon Hotel	0	1.30 p. m.	2.00 p. m.	7.00 a. m	7.30 a. m.
Arrive Tower Falls:					
Via Dunraven Pass	16	3.00 p. m.	3.45 p. m.	8.30 a. m.	9.15 a. m.
Via Mount Washburn	19	4.15 p. m	4.45 p. m.	9.45 a. m	10.15 a. m.
(For Cooke City Entrance see Note 1.)					
TOWER FALLS TO GARDINER.					
Leave Tower Falls	0	3.15 p. m.	4.45 p. m.	8.30 a. m.	10.15 a. m.
Arrive Mammoth Hot Springs	20	5.30 p. m.	6.45 p. m.	10.00 a. m.	12.15 p. m.
Leave Mammoth Hot Springs (via Main Road)	0	7.00 a. m.	7.30 a. m.	2.30 p. m.	3.00 p. m.
Arrive Gardiner Entrance	5	7.30 a. m.	8.00 a. m.	2.50 p. m	3.30 p. m.
MAMMOTH HOT SPRINGS TO GARDINER.					
Leave Mammoth Hot Springs (via Old Road)	0	8.45 a. m.	9.00 a. m	11.45 a. m.	1.00 p. m.
Arrive Gardiner Entrance	5	9.30 a. m.	9.45 a. m.	12.15 p. m.	1.45 p. m.

The Acting Superintendent of the park has authority to change these schedules if necessary

NOTE 1.—Owing to scarcity of travel on the roads named, automobiles will be permitted to travel without schedule on the roads between the South Entrance and the Thumb; between the East Entrance and the Lake; between the Northeast or Cooke City Entrance and Tower Falls Station; and between the West Entrance (Yellowstone, Montana), and the Northwest or Gallatin Station Entrance. Upon entering the main roads at the Thumb, Lake, Tower Falls, and the West Entrance, however, automobiles must conform to the regular schedules.

NOTE 2.—Automobiles making the morning trip from the Lake to the Canyon will be permitted to make the side trip to Artist Point, provided they keep within the schedule upon passing Canyon Soldier Station

NOTE 3.—The road from the Wylie Swan Lake Camp to Norris; the Norris-Fountain-Upper Basin-Thumb-Lake-Canyon-Norris road (called the Belt Line); and the road from the Canyon to Mammoth Hot Springs via Dunraven Pass, are open to automobile and truck travel without schedule from 6.30 p. m. to 6.45 a. m.

Approved:

STEPHEN T. MATHER,
Assistant to the Secretary.

R. B. MARSHALL,
Superintendent of National Parks.

Copyright by Haynes, St. Paul.

SOLDIER STATIONS ARE LOCATED AT STRATEGIC POINTS
THROUGHOUT THE PARK.

RULES AND REGULATIONS.

REGULATIONS APPROVED MAY 27, 1911.

The following rules and regulations for the government of the Yellowstone National Park are hereby established and made public pursuant to authority conferred by section 2475, Revised Statutes, United States, and the act of Congress approved May 7, 1894:

1. It is forbidden to remove or injure the sediments or incrustations around the geysers, hot springs, or steam vents; or to deface the same by written inscriptions or otherwise; or to throw any substance into the springs or geyser vents; or to injure or disturb in any manner or to carry off any of the mineral deposits, specimens, natural curiosities, or wonders within the park.

2. It is forbidden to ride or drive upon any of the geyser or hot-spring formations, or to turn stock loose to graze in their vicinity.

3. It is forbidden to cut or injure any growing timber. Camping parties will be allowed to use dead or fallen timber for fuel. When felling timber for fuel, or for building purposes when duly authorized, stumps must not be left higher than 12 inches from the ground.

4. Fires shall be lighted only when necessary, and completely extinguished when not longer required. The utmost care must be exercised at all times to avoid setting fire to the timber and grass.

5. Hunting or killing, wounding, or capturing any bird or wild animal, except dangerous animals when necessary to prevent them from destroying life or inflicting an injury, is prohibited. The outfits, including guns, traps, teams, horses, or means of transportation used by persons engaged in hunting, killing, trapping, ensnaring, or capturing such birds or wild animals, or in possession of game killed in the park under other circumstances than prescribed above, will be forfeited to the United States, except in cases where it is shown by satisfactory evidence that the outfit is not the property of the person or persons violating this regulation, and the actual owner thereof was not a party to such violation. Firearms will only be permitted in the park on written permission from the superintendent thereof. On arrival at the first station of the park, guard parties having firearms, traps, nets, seines, or explosives will turn them over to the sergeant in charge of the station, taking his receipt for them. They will be returned to the owners on leaving the park.

6. Fishing with nets, seines, traps, or by the use of drugs or explosives, or in any other way than with hook and line, is prohibited. Fishing for purposes of merchandise or profit is forbidden. Fishing may be prohibited by order of the superintendent of the park in any

RULES AND REGULATIONS—Continued.

of the waters of the park, or limited therein to any specified season of the year, until otherwise ordered by the Secretary of the Interior.

7. No person will be permitted to reside permanently or to engage in any business in the park without permission, in writing, from the Department of the Interior. The superintendent may grant authority to competent persons to act as guides and revoke the same in his discretion, and no pack trains shall be allowed in the park unless in charge of a duly registered guide.

8. The herding or grazing of loose stock or cattle of any kind within the park, as well as the driving of such stock or cattle over the roads of the park, is strictly forbidden, except in such cases where authority therefor is granted by the Secretary of the Interior. It is forbidden to cut hay within the boundaries of the park excepting for the use of the wild game and such other purposes as may be authorized by the Secretary of the Interior or the park superintendent.

9. No drinking saloon or barroom will be permitted within the limits of the park.

10. Private notices or advertisements shall not be posted or displayed within the park, except such as may be necessary for the convenience and guidance of the public, upon buildings on leased-ground.

11. Persons who render themselves obnoxious by disorderly conduct or bad behavior, or who violate any of the foregoing rules, will be summarily removed from the park, and will not be allowed to return without permission, in writing, from the Secretary of the Interior or the superintendent of the park.

12. It is forbidden to carve or write names or other things on any of the mileposts or signboards or any of the platforms, seats, railings, steps, or any structures or any tree in the park.

Any person who violates any of the foregoing regulations will be deemed guilty of a misdemeanor, and be subjected to a fine as provided by the act of Congress approved May 7, 1894, "to protect the birds and animals in Yellowstone National Park and to punish crimes in said park, and for other purposes," of not more than $1,000, or imprisonment not exceeding two years, or both, and be adjudged to pay all costs of the proceedings.

INSTRUCTIONS APPROVED APRIL 15, 1914.

1. The feeding, interference with, or molestation of any bear or other wild animal in the park in any way by any person not authorized by the superintendent is prohibited.

2. *Fires.*—The greatest care must be exercised to insure the complete extinction of all camp fires before they are abandoned. All

RULES AND REGULATIONS— Continued.

ashes and unburned bits of wood must, when practicable, be thoroughly soaked with water. Where fires are built in the neighborhood
of decayed logs, particular attention must be directed to the extin-
guishment of fires in the decaying mold. Fire may be extinguished
where water is not available by a complete covering of earth, well
packed down.

Especial care should be taken that no lighted match, cigar, or cigarette
is dropped in any grass, twigs, leaves, or tree mold.

3. *Camps.*—No camp will be made at a less distance than 100 feet
from any traveled road. Blankets, clothing, hammocks, or any other
article liable to frighten teams must not be hung at a nearer distance
than this to the road. The same rule applies to temporary stops, such
as for feeding horses or for taking luncheon.

Many successive parties camp on the same sites during the season,
and camp grounds must be thoroughly cleaned before they are abandoned. Tin cans must be flattened and, with bottles, cast-off clothing, and all other débris, must be deposited in a pit provided for the
purpose. When camps are made in unusual places, where pits may
not be provided, all refuse must be hidden where it will not be offen-
sive to the eye.

4. *Concessionaires.*—All persons, firms, or corporations holding
concessions in the park must keep the grounds used by them properly
policed and maintain the premises in a sanitary condition to the
satisfaction of the superintendent.

5. *Bicycles.*—The greatest care must be exercised by persons using
bicycles. On meeting a team the rider must stop and.stand at side of
road between the bicycle and the team—the outer side of the road if
on a grade or curve. In passing a team from the rear the rider
should learn from the driver if his horses are liable to frighten, in
which case the driver should halt and the rider dismount and walk
past, keeping between the bicycle and the team.

6. *Fishing.*—All fish less than 8 inches in length should at once be
returned to the water with the least damage possible to the fish. Fish
that are to be retained must be at once killed by a blow on the back
of the head or by thrusting a knife or other sharp instrument into
the head. No person shall catch more than 20 fish in one day.

7. *Dogs.*—Dogs are not permitted in the park.

8. *Grazing animals.*—Only animals actually in use for purposes of
transportation through the park may be grazed in the vicinity of the
camps. They will not be allowed to run over any of the formations
nor near to any of the geysers or hot springs; neither will they be
allowed to run loose within 100 feet of the roads.

9. *Formations.*—No person will be allowed on any formations after sunset without a guide.

10. *Hotels.*—All tourists traveling with the authorized transportation companies, whether holding hotel coupons or paying cash, are allowed the privilege of extending their visit in the park at any of the hotels without extra charge for transportation. However, 24 hours' notice must be given to the managers of the transportation companies for reservations in other coaches.

11. *Driving on roads of park.*—(a) Drivers of vehicles of any description, when overtaken by other vehicles traveling at a faster rate of speed, shall, if requested to do so, turn out and give the latter free and unobstructed passageway.

(b) Vehicles in passing each other must give full half of the roadway. This applies to freight outfits as well as any other.

(c) Racing on the park roads is strictly prohibited.

(d) Freight, baggage, and heavy camping outfits on sidehill grades throughout the park will take the outer side of the road while being passed by passenger vehicles in either direction.

(e) In making a temporary halt on the road for any purpose all teams and vehicles will be pulled to one side of the road far enough to leave a free and unobstructed passageway. No stops on the road for luncheon or for camp purposes will be permitted. A team attached to a vehicle will not be left without the custody of a person competent to control it; a team detached from a vehicle will be securely tied to a tree or other fixed object before being left alone.

(f) In rounding sharp curves on the roads, like that in the Golden Gate Canyon, where the view ahead is completely cut off, drivers will slow down to a walk. Traveling at night is prohibited except in cases of emergency.

(g) Transportation companies, freight and wood contractors, and all other parties and persons using the park roads will be held liable for violation of these instructions.

(h) Pack trains will be required to follow trails whenever practicable. During the tourist season, when traveling on the road and vehicles carrying passengers are met, or such vehicles overtake pack trains, the pack train must move off the road not less than 100 feet and await the passage of the vehicle.

(i) During the tourist season pack animals, loose animals, or saddle horses, except those ridden by duly authorized persons on patrol or other public duties, are not permitted on the coach road between Gardiner and Fort Yellowstone.

(k) Riding at a gait faster than a slow trot on the plateaus near

RULES AND REGULATIONS—Continued.

the hotels where tourists and other persons are accustomed to walk is prohibited.

(*l*) Mounted men on meeting a passenger team on a grade will halt on the outer side until the team passes. When approaching a passenger team from the rear, warning must be given, and no faster gait will be taken than is necessary to make the passage, and if on a grade the passage will be on the outer side. A passenger team must not be passed on a dangerous grade.

(*m*) All wagons used in hauling heavy freight over the park roads must have tires not less than 4 inches in width. This order does not apply to express freight hauled in light spring wagons with single teams.

12. *Liquors.*—All beer, wine, liquors, whisky, etc., brought into the Yellowstone National Park via Gardiner to be carried over the roads through the reservation to Cooke City, must be in sealed containers or packages, which must not be broken in transit.

13. *Miscellaneous.*

Persons are not allowed to bathe near any of the regularly traveled roads in the park without suitable bathing clothes.

14. *Penalty.*—The penalty for disregard of these instructions is summary ejection from the park.

Notices.—(*a*) Boat trip on Yellowstone Lake: The excursion boat on Yellowstone Lake plying between the Lake Hotel and the Thumb lunch station at the West Bay is not a part of the regular transportation of the park, and an extra charge is made by the boat company for this service.

(*b*) Side trips in park: Information relative to side trips in the park and the cost thereof can be procured from those authorized to transport passengers through or to provide for camping parties in the park; also at the office of the superintendent.

(*c*) All complaints by tourists and others as to service, etc., rendered in the reservation should be made to the superintendent in writing.

OTHER NATIONAL PARKS.

The circulars containing information about National Parks listed below may be obtained free of charge by writing to the Secretary of the Interior, Washington, D. C.

Yosemite National Park.
Mount Rainier National Park.
Crater Lake National Park.
Mesa Verde National Park.

Sequoia and General Grant National Parks.
The Hot Springs of Arkansas.
Glacier National Park.

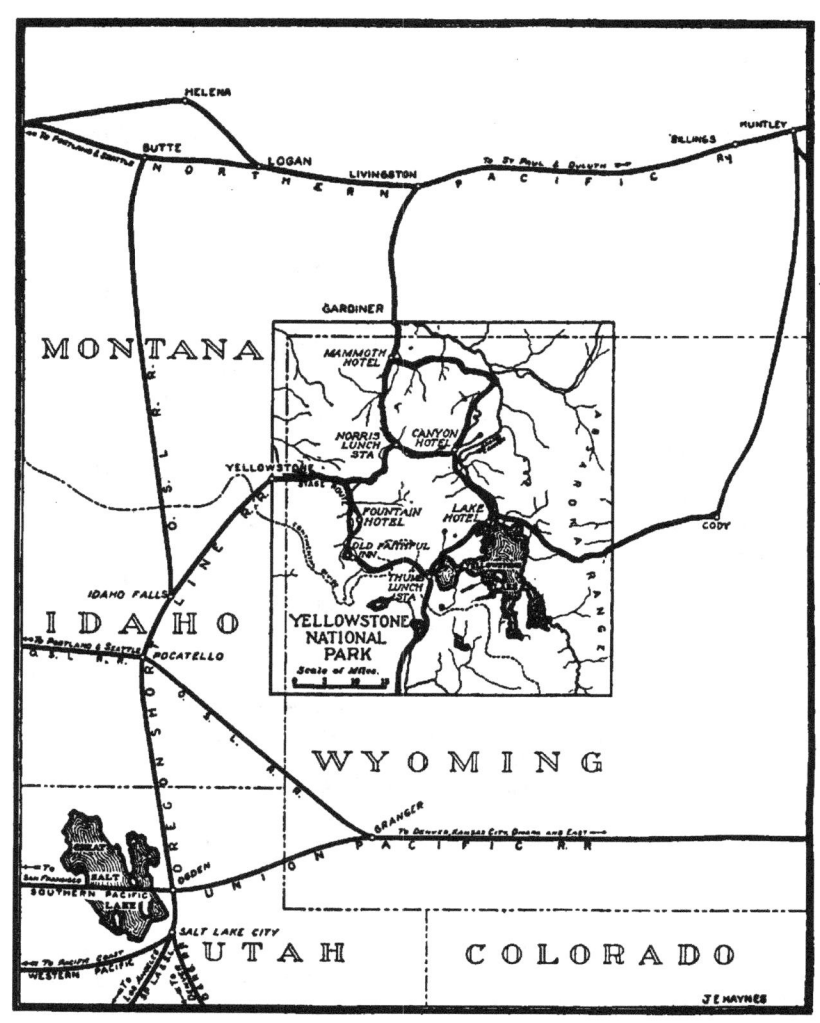

RAILWAY APPROACHES TO YELLOWSTONE NATIONAL PARK.
(The Burlington Route has a terminus at Cody, Wyo.)

IMPORTANT.

For a description of the objects of interest in the park in the order in which you will see them on this trip begin on the page indicated for your entrance:

Northern Entrance, page **113.**

(Reached by Northern Pacific Railway to Gardiner, Mont.)

Western Entrance, page **25.**

(Reached by the Union Pacific System (O. S. L. R. R.) at Yellowstone, Mont.)

Eastern Entrance, page **88.**

(Reached by roadway from Cody, Wyo., on the Burlington Route.)

Southern Entrance, page **85.**

(Reached by wagon road from Jackson's Hole, Wyo.)

THE LARGEST TROUT EVER CAUGHT IN THE PARK.

Caught in 1914 with a bamboo rod and Colorado Spinner.

Weight—18½ pounds.

Length—39½ inches.

Place—Shoshone Lake.

TOUR OF THE PARK
FROM THE WESTERN ENTRANCE

Yellowstone, Montana.—On November 12, 1907, the Oregon Short Line R. R. completed its branch line to Yellowstone, twenty miles from the Lower Geyser Basin, where it has constructed a unique stone depot having every convenience for the traveler; private dressing rooms, check rooms where extra baggage may be stored while making the park trip, and a covered porte-cochere and loading platform. Coats, hats and linen dusters are for rent for the park trip. The dining car department of this railroad operates an eating-house nearby where exceptionally good meals are served.

Copyright by Haynes, St. Paul.

YELLOWSTONE STATION, OREGON SHORT LINE R. R.

Tourists making the hotel trip through the park are transported in elegant four-horse Concord coaches of the Yellowstone-Western Stage Company to all points in the reserve.

Christmas Tree Park, a beautiful plateau, extends for ten miles along the western boundary and is about three miles wide where the road crosses it. The government engineers constructed an ideal roadway here which has a bed of crushed rock and an oiled surface for several miles.

The drive from Yellowstone to the Fountain Hotel is up the Madison River past Mt. Burley, National Park Mt., and the Cascades of the Firehole. This is

CHRISTMAS TREE PARK AT WESTERN ENTRANCE.

"NO PERSON SHALL CATCH MORE THAN TWENTY FISH IN ONE DAY."

the route which was the pioneer entrance to Yellowstone Park; having been used by the early explorer James Bridger, discoverer of the Great Salt Lake, Colter of the Lewis and Clark expedition, and Dr. F. V. Hayden of the U. S. Geological Survey.

After leaving Christmas Tree Park the beautiful Madison River comes into view; a Wylie Camp is situated a short distance from this point and farther on, the Riverside Military Station headquarters for a detachment of United States Cavalry.

The Rainbow and Loch Leven Trout of the Madison River have made this section of the park famous. It is not uncommon for an expert angler to land a six-pound rainbow trout in this vicinity, a sport to be fully appreciated only by experience. The United States Fish Commission's work in the Yellowstone reserve as a whole is to be commended, many ideal streams having been destitute of fish life before being stocked.

Mt. Burley rises from the water's edge several hundred feet high on the south side of the Madison

NATIONAL PARK MOUNTAIN

Canyon, a rugged escarpment of lava rock. The scenery in Madison Canyon is acknowledged by all to be equalled only by the Grand Canyon of the Yellowstone.

National Park Mountain, situated at the junction of the Gibbon and Firehole Rivers, marks the point where on September 19, 1870, the Washburn-Langford Party camped after having completed the exhaustive

JUNCTION
OF THE
GIBBON and FIREHOLE RIVERS
Forming the MADISON Fork of the MISSOURI
On the Point of Land Between the Two Tributary Streams
SEPTEMBER 19 1870 the WASHBURN EXPEDITION Which
First Made Known to the World the Wonders of Yellowstone
Was Encamped and Here Was First Suggested the Idea
Of Setting Apart This Region as A NATIONAL PARK

To FOUNTAIN 7½ MI.

To NORRIS BASIN 14 MI.

SIGN NEAR NATIONAL PARK MOUNTAIN AT THE JUNCTION
OF THE ROADS.
Left road to Norris Basin.   Right road to Lower (Fountain) Basin.

exploration of the park. While encamped at this place one member of the party suggested that the Yellowstone region, just explored, should be made a National Park, and it was largely through their efforts that Congress in 1872 passed the act of dedication.

One of the most charming drives in the reservation is from National Park Mountain to the **Cascades of the Firehole.** Here a short halt is usually made so these beautiful cascades may be viewed from different points. Below the upper cascades the river is confined in a narrow gorge until it reaches the main falls. The Firehole River owes a large part of its flow to the immense drainage from the geyser basins, and in many places along its course the water is appreciably warm; in spite of this fact, however, trout abound in its pools all the way from Madison Lake, its source, to these cascades.

After passing the second Military Station about two miles from the Lower Geyser Basin, Nez Percé Creek, made famous by the Nez Percé Indians headed by Chief Joseph on their memorable raid through the park in 1877, is passed. **The Shaw & Powell Nez Perce Camp** is situated here.

Lower Geyser Basin.—This is a comparatively wide valley, extending southward from the junction of the east fork of Firehole River with the main stream, and embracing an area of thirty or forty square miles. In this valley Dr. Hayden, in his official survey of the park region has catalogued 693 hot springs.

CHIEF JOSEPH OF THE NEZ PERCÉ INDIANS.

The surrounding heavily-timbered slopes rise 400 to 800 feet above the geyser basin.

The Fountain Hotel (Alt. 7,180 feet) is pleasantly situated on the east side of the valley, commanding an extended view of the surroundings. Its appointments include electric light and steam heat, and it is the only hotel in the park having natural hot water baths. It is the first hotel reached by visitors entering the park from the west. The nearby streams are stocked with "Loch Leven" and "Eastern Brook" trout.

The chief attractions at the Lower Geyser Basin are the Fountain and Great Fountain Geysers, the Mammoth Paint Pots, Clepsydra Spring and Firehole Lake.

Fountain Geyser, about 2,000 feet south of the hotel, occupies an eminence built up by its own deposit over an area of several acres.

GEYSER TABLE.

LOWER AND MIDWAY BASINS.

Corrected by observations made during the past season.

Geysers at LOWER BASIN	Maximum Height	Duration	Intervals of Eruption
Fountain	75 ft.	20 min.	irregular
Great Fountain	100 ft.	30 min.	8 to 12 hours
AT MIDWAY BASIN			
Excelsior	300 ft.	Variable	1 to 4 hrs. Ceased to play in 1888.

This geyser has proven very unreliable the past few seasons, but during periods of activity indications of eruptions are as follows: When both the pool and crater are full of water to the rim, it is probable that an eruption will soon take place. Immediately after action the water falls from twelve to eighteen inches below the crater rim, from which point it spouts gradually until the climax is reached.

In July, 1899, the Fountain Geyser ceased operations and remained inactive until October, when it resumed

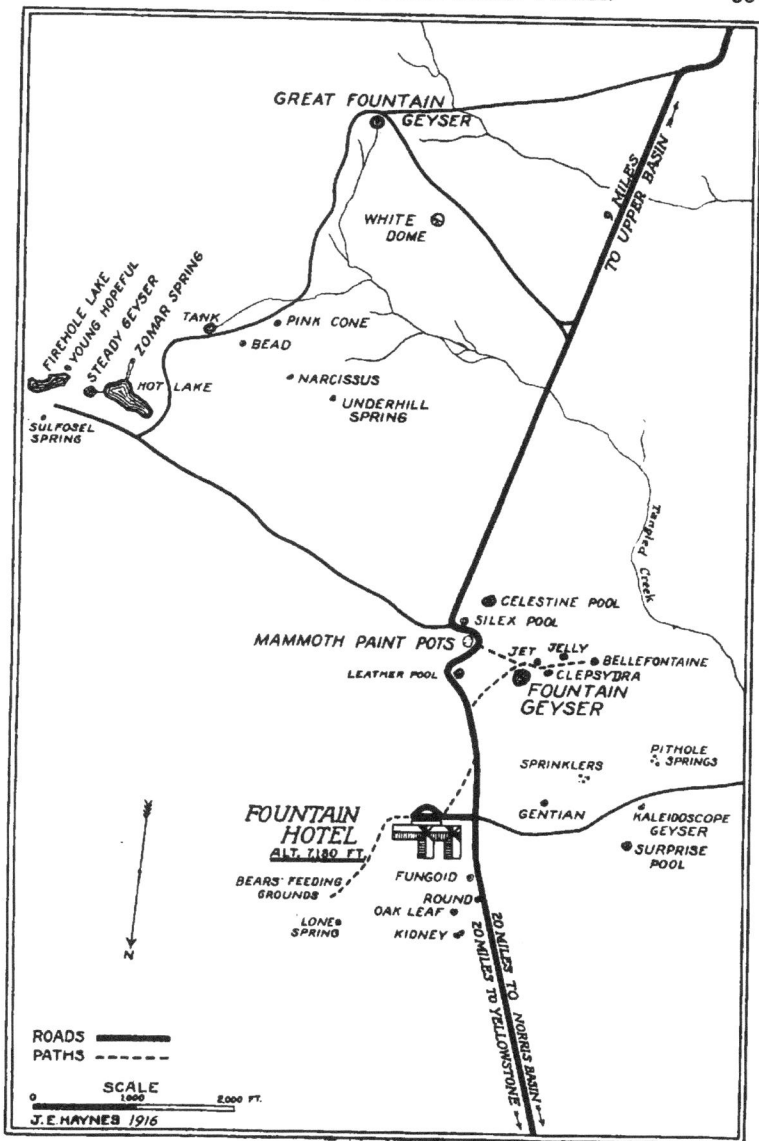

LOWER GEYSER BASIN.

CUB BEARS CAUGHT NEAR THE FOUNTAIN HOTEL AND SHIPPED TO AN
EASTERN ZOOLOGICAL GARDEN.

its usual displays. In the meantime an immense geyser broke out in the large pool north of the Fountain.
Its eruptions were of great force, quite irregular, but
equal to those of Old Faithful, and continuing, at
times, fully an hour.

In July, 1909, ten years later, it abandoned its crater
for the one adjoining and threw out jagged masses of
geyserite more than 200 feet. The water was muddy
and full of rock fragments for many hours; and as
late as September, large pieces of rock were thrown
out during the more violent eruptions.

For two days preceding the breaking out of this
geyser in its new place, much disturbance was noted
in the vicinity; loud rumblings were heard and the

thumping of the entombed steam and water, gaining in violence each hour, alarmed even those most used to the strange phenomena of the geyser region. During the remainder of the season of 1909 the Fountain Geyser played much higher than before, like a stream through a smaller nozzle, but its eruptions were less regular.

Clepsydra Spring, some fifty feet west from the Fountain, has recently developed into an active geyser of no small eruptive power, its frequent displays being really quite violent for so small a "spouter" and very pleasing withal.

Mammoth Paint Pots.—Some few hundred feet east of the Fountain, near the road, from which they are

MAMMOTH PAINT POTS, LOWER BASIN.

separated by a fringe of trees, are situated these wonderful paint pots. This remarkable mud caldron has
a basin which measures 40x60 feet with a mud rim on
three sides, which is from four to five feet in height.
In this basin is a mass of fine, whitish substance which
is in a state of constant agitation. It resembles some
vast boiling pot of paint or bed of mortar with numerous points of ebullition; and the constant boiling has
reduced the contents to a thoroughly mixed mass of
silicious clay. There is a continuous bubbling up of
mud, which, rising in hemispherical masses, cones,
rings and jets, produces sounds like a whispered
"plop-plop."

MAMMOTH PAINT POTS.

Great Fountain Geyser is situated about two miles south of the hotel and about one mile east of the main road; the one to the geyser branches off soon after crossing Fountain Creek. The description by Mr. David E. Folsom, who witnessed a display October 1, 1869, faithfully portrays its present exhibitions. "The hole through which the water was discharged was ten feet in diameter, and was situated in the center of a large circular shallow basin into which the water fell. There was a stiff breeze blowing at the time, and by going to the windward side and carefully picking our way over convenient stones we were enabled to reach the edge of the hole. At that moment the escaping steam was causing the water to boil up in a fountain five or six feet high. It stopped in an instant, and commenced settling down—twenty, thirty, forty feet—until we concluded that the bottom had fallen out, but the next instant, without any warning, it came rushing up and shot into the air at least eighty feet, causing us to stampede. It continued to spout at intervals of a few moments for some time, but finally subsided." Many interesting and curious sights in the vicinity of the Great Fountain should be visited. The "White Dome," "Surprise," **Firehole Lake,** "Mushroom," and **Buffalo Spring** are the most prominent. The last was discovered in 1869 by an early exploring party. In describing their trip the writer says: "In one of these springs we saw the whitened skeleton of a mountain buffalo that had probably fallen in accidentally. No king or saint was ever more magnificently entombed than this monarch of the hills in his sepulchre in the wilderness."

Midway Geyser Basin.—Strictly speaking, this section constitutes the upper portion of the Lower Basin, and is three miles from the Lower Basin. Being about midway between the extremes of the Upper and Lower Geyser Basins, this locality is given a distinct designation.

Excelsior Geyser.—"Early explorers in this locality discovered, in 1871," says Dr. Peal, "on the west bank of Firehole River, an immense pit of rather irregular outline, 330 feet in length by 200 feet in width at the widest part. The water is of a deep blue tint, and is intensely agitated all the time, dense clouds of steam

EXCELSIOR GEYSER, MIDWAY BASIN, WHICH CEASED TO PLAY IN 1888.

constantly ascending from it. It is only when the breeze wafts this aside that the surface of the water, which is fifteen or twenty feet below the level surrounding, can be seen. The walls on three sides are perpendicular, cliff-like, and in places overhang, having been worn away on the other." Visited by thousands annually, this section became known as "Hell's Half Acre," a name it retained until 1881, when discovered by Colonel P. W. Norris to be a geyser of great force, and then named by him "Excelsior." Its eruptions in 1881 began after the tourist season had closed; Colonel Norris witnessed thirty eruptions, varying from 75 to 250 feet in height, at intervals of one to four hours. The intervals of eruption during 1888 were at first about every hour and fifteen minutes, increasing towards the latter part of the season to two hours. Immediately preceding each eruption a violent upheaval occurred, raising the entire volume of water in the crater nearly fifty feet, then instantly one or two, and sometimes three, terrific explosions would occur, followed closely by the shooting upwards of columns of water, and oftentimes masses of the rocky formation, to a height of 200 to 250 feet. Tons of rock have in this way been hurled into the Firehole River, some pieces falling 500 feet from the crater. At each upheaval sufficient water would escape to raise the river several inches. The Excelsior Geyser ejected as much water at each eruption, as all the other geysers combined.

Turquoise Spring, situated about 150 feet north of Excelsior, is a silent pool, about 100 feet in diameter, and remarkable for its beautiful blue transparent water. There is a constant overflow from the spring, through a shallow channel some two feet wide, its sides and bottom being exquisitely colored; when Excelsior was in action the water in this spring sank fully ten feet and did not resume its normal condition for nearly a year. West of the Turquoise Spring, and in itself a marvel, is a small spring of cold water, which, though rather "brackish" to be palatable, is attractive as being the sole cold spring in this region of thermal waters.

Prismatic Lake is probably the very largest and certainly one of the most beautiful springs in the entire Park region. It is situated some 500 feet west of Excelsior Geyser, its dimensions being 250x400 feet. Over the central pit, or bowl, of this spring the water is of a deep blue color, changing to green towards the margin, while that in the shallower portions of the lake surrounding the central basin has a yellow tint gradually fading into orange. Outside its rim there is a brilliant red deposit, which shades into purples, browns, and grays, all seemingly painted upon a ground of grayish white, which forms the mound, built up of layers of silicious deposit, upon which the spring is situated. This coloring is in vivid bands, which are strikingly marked and distinct. The water flowing off in every direction, with constant wave-like pulsations over the artistically scalloped and slightly raised rim of the lake, has formed a succession of terraces, each a few inches in height, down the slopes of the mound, particularly upon its southern face. It is impossible to exaggerate the delicacy and richness of the coloring in and about this wonderful phenomenon of nature. The temperature of the water is about 146 degrees

PRISMATIC LAKE, MIDWAY BASIN.

Fahrenheit. The constantly rising clouds of steam
sometimes render difficult a good view of the lake surface; but viewed from the proper standpoint (generally
with the sun to the back), these same volumes of steam
are exceedingly attractive, reflecting the colors of the
rainbow or prism, though some attribute it to the var-
iegated tints of its waters. The entire drive from Midway to the Upper Basin, some five miles, is among
many natural wonders.

Biscuit Basin is on the west side of Firehole River
and on the north side of Iron Spring Creek, being about
one mile below Riverside Bridge. In Biscuit Basin is
Sapphire Pool, whose highly ornamented margin suggested the basin's rather odd name. Hundreds of small
symmetrical, biscuit-like knobs of olive-green formation surround the spring, which is of the variety known
as pulsating or breathing springs (geysers in fact).
The constant ebb and flow of its waters have produced
this peculiar formation, from one to another of which
one must pick one's way in order to get a good view of
the pool itself. A few feet to the west is

Jewel Geyser, whose eruptions occur with the remarkable frequency of from three to five minutes,
throwing jets of water to a height of about forty
feet. Scarce 500 feet further west are the **Black Pearl**
and **Silver Globe.** The former has a beautiful basin,
studded thickly with black pearls, each about one-
quarter of an inch in size. A curious feature of this
little "spouter" is the fact that its formation surrounds
the roots and stump of a tree, completely incrusting
the same with its rich, black ornamentations.

The Silver Globe derives its name from the constant
rising to its surface of large, silvery globules or bubbles
of gas or steam, which, of course, immediately disappear on reaching the air.

Artemisia Geyser is situated between the road and

A PRIVATE CAMPING PARTY.

the river, quite near the former, which is elevated some
twenty feet above the spring. Stepping to the edge
of the bank, an excellent view of the crater is obtained,
the crystal clearness of its waters allowing a distinct
view in to its apparently bottomless depths. The spring
is sixty feet in diameter and generally very little agitated, merely overflowing. The surrounding formation,
quite unlike that of any other spring or geyser, is as

hard as flint, and of a peculiar olive-green color. Although for the most part very quiescent, this spring has occasional pulsations in the nature of eruptions, at which times large quantities of water are forced out, fairly flooding the formation between it and the river.

"HIKERS" AND THEIR PATIENT BEAST OF BURDEN.

These eruptions occur at intervals of twelve to twenty-four hours. The bank of the Firehole, some thirty feet high at this point, is the most highly colored section of the river to be found in the Upper Basin. The best view is obtained from the bridle-path on the opposite side of the river. This trail leads south from the Splendid, crossing the Firehole just above its confluence with Iron Spring Creek, near which it joins the main road.

Morning Glory Spring is passed just before coming to the Riverside Bridge. The symmetrical shape and funnel-like crater whose walls are delicately colored, account for the appropriate name of this spring. At the surface the diameter is 23 feet and the temperature 100 degrees F., and apparent depth 29 feet.

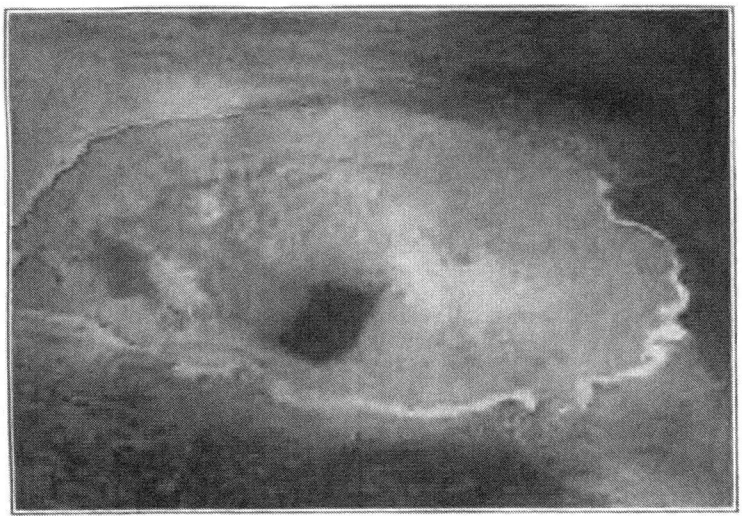

MORNING GLORY SPRING, UPPER BASIN.

Upper Geyser Basin is triangular in form and embraces an area of about four square miles; it contains twenty-six geysers and upwards of 400 hot springs Iron Spring Creek bounds it on the west; timbered mountain slopes form the hypotenuse of the triangle and a wavy line of dark forest conifers, its southern base. The main Firehole River drains it, centrally; its shelving banks are thickly pitted with steaming hot springs and studded with mounds and cones of geyser-

GEYSER TABLE.
UPPER GEYSER BASIN.

Corrected by observations made during the past season.

Geysers at UPPER GEYSER BASIN	Maximum Height	Duration	Intervals
Artemisia	30 ft.	10 min.	12-24 hrs.
Beehive	200 ft.	8 min.	8 hrs. to 8 days.
Castle	75 ft.	30 min.	26 hrs. Frequently misses.
Cliff	100 ft.	8 min.	4-8 hrs.
Cub (Big)	30 ft.	10 min.	With Lioness.
Cub (Little)	6 ft.	3 min.	Frequently.
Daisy	75 ft.	3 min.	1½ hrs. to 1 hr. and 50 min.
Economic	20 ft.	10 sec.	5 min. to 2 hrs.
Fan	6 ft.	10 min.	4-6 hrs.
Giant	250 ft.	1½ hrs.	7-12 days.
Giantess	100 ft.	12-24 hrs.	4-12 days.
Grand	200 ft.	30-60 min.	1-10 days.
Grotto	30 ft.	15 min-8 hrs.	2-8 hrs.
Jewel	30 ft.	2 min.	5 min.
Lion	60 ft.	3 min.	2-8 hrs.
Lioness	100 ft.	10 min.	15-20 days.
Lone Star	50 ft.	10 min.	1-2 hrs.
Mortar	30 ft.	5 min.	2 hrs.
Oblong	20 ft.	5 min.	7-8 hrs.
Old Faithful	150 ft.	4 min.	65-75 min.
Riverside	100 ft.	20 min.	6-7 hrs.
Rocket	50 ft.	2-3 min.	2-8 hrs.
Sawmill	40 ft.	2 hrs.	3-4 hrs.
Spasmodic	10 ft.	10 min.	2-3 hrs.
Splendid	200 ft.		Inactive since 1892.
Sponge	4 ft.	15 sec.	1-¼ min.
Turban	25 ft.	20-60 min.	With Grand & frequently.

The siren at the Haynes Picture Shop announces the playing of the larger geysers.

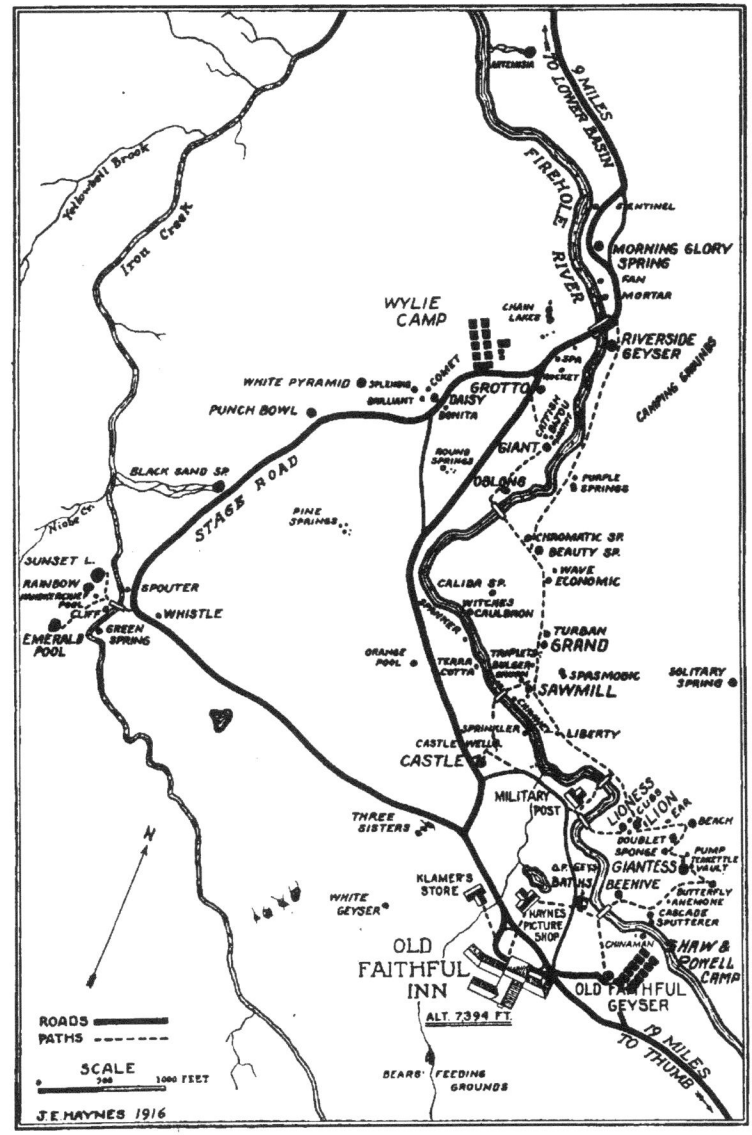

UPPER GEYSER BASIN.

ite. Here, grouped within the narrow space of perhaps a square mile are the grandest and mightiest geysers known to man; and silent pools of scalding, meteoric water that for beauty of formation and delicacy of coloring are marvels. The surface of the basin consists, for the most part, of a succession of gentle undulations, each crowned with a geyser-cone or hot-spring vent

HOWARD EATON, WOLF, WYOMING,
A NOTED YELLOWSTONE PARK GUIDE.

and covered with layers of silicious sinter that give it a grayish-white, sepulchral hue. Clouds of vapor hang shroud-like above it; the earth trembles and is filled with strange rumblings, the air is heavy with sulphurous fumes, and vegetable life is extinct. In a paper read before the Cardiff (Wales) Naturalists' Society, Mr. Charles T. Whitmell said: "Nowhere else, I believe, can be seen, on so grand a scale, such clear evidence of dying volcanic action. We seem to witness the death throes of some great American Enceladus. Could Dante have seen this region, he might have added another terror to his Inferno."

The **Fan and Mortar Geysers** are near the river between Morning Glory Spring and the Riverside Gey-

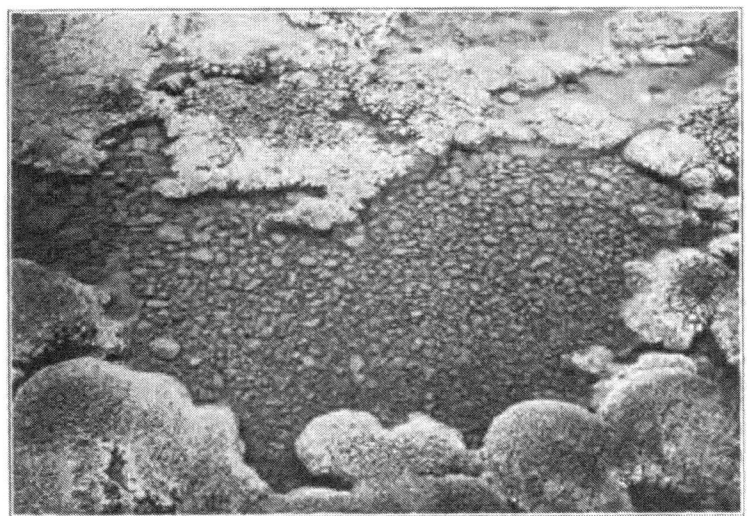

GEYSER EGGS AT SAWMILL GEYSER.
(One-tenth actual size.)

ser, about three hundred feet down stream from the latter. Intervals between eruptions of the Fan vary from four to six hours; it plays for ten minutes but only six or eight feet high. The Mortar plays thirty feet high for five minutes every two hours.

The **Riverside Geyser,** which is situated on the east bank of the Firehole River a few feet above the new steel bridge, erupts every six or seven hours, obliquely across the river; sometimes eruptions take place as frequently as each five and one-half hours for a period of several days, probably on account of an increased supply of water.

The Riverside formation is made up of two craters

Copyright by Haynes, St. Paul.
GROTTO GEYSER FORMATION.

on a chimney-like mound of silicious deposit; the lower, or main crater, overflows continuously for about an hour-before each eruption; jets of water are thrown out about twenty minutes before displays, from the upper crater. The maximum height of the Riverside is one hundred feet; this is maintained for eight minutes, followed by the characteristic steam-period lasting several minutes.

The next feature of prominence passed is the **Grotto Geyser,** which has the most extraordinary formation of any geyser in the park; it received this appropriate name in 1870 from the Washburn party. Eruptions vary in interval from two to eight hours, and are about thirty feet high, lasting any length of time from fifteen

GIANT GEYSER, UPPER BASIN.

minutes to eight hours. Occasionally the Grotto ceases and the **Rocket,** an isolated cone a few feet north, plays to a height of fifty feet for two or three minutes; then the Grotto resumes activity. The pool near the road eighty feet north of the Rocket is called the **Spa** (a mineral spring); it has not been observed to erupt, but empties and fills at intervals indicating a probable relation to some distant geyser.

Copyright by Haynes, St. Paul.

COACHES UNLOADING AT WYLIE CAMP.

The **Giant Geyser,** situated about five hundred feet southeast of the Grotto, is the *highest geyser in the world*; it plays two hundred and fifty feet, for a period of one and one-half hours, every seven to twelve days. Its maximum height, however, is maintained only during the first twenty minutes. The Giant Geyser cone is ten feet high and has one side partly broken off, exposing to view its channel, which is four feet across.

On the same platform of deposit with the Giant are three boiling cauldrons, the **Catfish, Bijou** and **Mastiff,** all of minor importance. Near these is a sign marked "Indicator," but it is very uncertain if activity of the Giant is ever foretold by activity of these smaller basins. In some cases however, geysers do have true indicators, notably the Beehive.

Copyright by Haynes, St. Paul.

THE UPPER BASIN CAMP OF THE WYLIE PERMANENT CAMPING
COMPANY.

The **Daisy Geyser,** located near the **Wylie Upper Basin Camp,** and the **White Pyramid,** is a very pretty and reliable geyser. The character of its eruptions, which occur every one and one-half or two hours, are very like the **Splendid Geyser** which ceased to play about the time the Daisy broke out in 1892. The Daisy plays seventy-five feet high; duration, three minutes. Across the road from the Daisy is **Bonita Pool,** which acts as its indicator. The **Brilliant** is a beautiful, blue, quiescent hot spring. Near it is the **Comet,** called a geyser in the past, but now inactive; it still boils up at intervals, and has built up a small cone of geyserite.

Punch Bowl Spring.—The wagon road leading westward from the Splendid toward Black Sand and Sunset Basins passes the Punch Bowl, by far the handsomest spring of its peculiar class to be found in the geyser

Copyright by Haynes, St. Paul.

DAISY GEYSER—75 FT.

PUNCH BOWL SPRING, UPPER BASIN.

region, if not in the world. Situated on the summit
of a small mound of silicious deposit, some five feet
above the general level, it is about ten feet in diameter,
with a glittering rim of brilliantly colored formation
eighteen inches in height. The constant boiling of its
contents, though only a small part of its surface is agitated, as the bubbles of escaping steam and gas arise,
produces a wáve-like undulation over the entire spring
and gives it a steady and not inconsiderable overflow.
A small cave-like opening on the east side of the mound
or cone is very handsome, having the appearance of
being lined with satin of the rarest beauty and texture.
Early visitors to the Park during the seasons of 1873
and 1875 speak of this spring as being an active geyser,

and during 1888 similar reports gained currency. Nothing, however, is certainly known as to the correctness of these reports.

Black Sand Spring and Specimen Lake.—Dr. Peale's description of Black Sand Spring is interestingly comprehensive, and is as follows: "This is one of the most beautiful springs in the Upper Basin. It has a delicate rim, with toadstool-like masses around it. The basin slopes rather gently toward a central aperture that, to the eye, appears to have no bottom. The water in the spring has a delicate turquoise tint, and as the breeze sweeps across its surface, dispelling the steam, the effect of the ripple of the water is very beautiful. The sloping sides are covered with a light brown crust; sometimes it is rather a cream color. The funnel is about forty feet in diameter, while the entire space covered by the spring is about 55x60 feet, outside the rim of which is a border of pitch-stone (obsidian) sand or gravel sloping twenty-five feet. From its west side flows a considerable stream, forming a most beautiful channel, in which the coloring presents a remarkable variety of shades; the extremely delicate pinks are mingled with equally delicate tints of saffron and yellow, and here and there shades of green." The overflow from this spring spreads out over a large area, called **Specimen Lake,** which deserves more than passing notice. Absorption of the surrounding silica has destroyed many of the trees in the vicinity, the dry, lifeless trunks adding to the attractiveness of the place, geologically speaking, by affording the appearance of petrifactions.

After Black Sand Spring, the next attractions are Sunset Lake and Emerald Pool, reached by a foot-bridge over Iron Spring Creek.

Sunset Lake is a beautifully colored pool which steams constantly and, though always boiling hot, never erupts. It is larger than **Rainbow Pool** and situated a few steps north of it. Both are very beautiful, though usually completely enveloped in steam. Several yards north at the edge of the timber is the most beautiful pool in the Upper Basin, **Emerald Pool**; its deep emerald color blends to yellow toward the edge, and the formation is a rich red immediately around it. This pool, though hot, never boils, and is slightly overflowing. Across the river from Emerald Pool is **Green Spring.**

Handkerchief Pool is but a few feet from Rainbow Pool, a small basin with a funnel-shaped opening. A handkerchief placed in the water near the edge will be drawn downward and out of sight by convection currents in the water, and in a few minutes will re-appear.

Cliff Spring usually is boiling violently; and though credited by some with having occasional eruptions, it is usually considered to be only a spring. It is close to the foot-bridge on the west side of the river.

The **Whistle,** situated near the road leading toward Old Faithful Inn, performs only at great intervals; but when the great rush of steam commences, as it does several times each season, a whistle-like roar is produced which is audible half a mile and lasts several minutes.

The **Three Sisters** springs, while attractive, are so like a hundred other boiling pools that they are usually passed without a halt. They are situated in sight of Old Faithful Inn and not far from the Castle Geyser (on the road leading direct from the Riverside Geyser to the hotel).

The **Castle Geyser** is at once recognized by its large cone resembling "an old feudal castle partially in ruins" (Doane). It occupies a prominent position and is visible from the hotel and nearly all parts of the basin. The great amount of deposit, perhaps 100 feet in diameter at its base, and the possession of the largest cone in the whole region, while giving it

CASTLE GEYSER, UPPER BASIN.

an air of conspicuousness at the same time indicate that it is one of the oldest active geysers in the Park. The broken condition of its cone on the east side renders possible an easy ascent to its summit, which is about twenty feet across. The orifice of the geyser tube in the top of the cone is about three feet in diameter, quite round, and is lined with a formation of bright orange color. Eruptions of the Castle occur at intervals of about twenty-six hours, preceded by the occasional throwing out of jets of water to the height of fifteen or twenty feet, perhaps. These premonitory symptoms of eruption generally continue five or six hours, when more violent demonstrations, during which columns of water are shot upward to a height of fully seventy-five feet, ensue, and, continuing for half an hour or so, are followed by a "steam period" similar to that of the Giantess. Several times each season it has eruptions of an unusual character, in which its columns of water are thrown to twice their usual height and its subsequent "steam periods" are proportionately forcible. A violent boiling spring is situated near the base of its cone, on the north side, which used to be a favorite resort of the "camper-out" in earlier days. It is ten feet across, has an apparent depth of 52 feet and a temperature of 199 degrees F.

Castle Well, a large, crested spring 100 feet north from the Castle, is usually very handsome. It generally is filled to overflowing, and the bottom and edges of the channel leading out of the north side are very highly and beautifully colored. This spring is twenty feet in diameter and overflows on two sides.

Old Faithful Inn (Alt. 7,394 feet), the most extensive log structure yet devised by man, with every convenience and luxury of the modern hotel, is the latest triumph in utilizing primitive material in constructing so unique a building. The rough blocks of stone which form its foundation appear as natural as when found at the base of the cliffs of the surrounding mountains.

OLD FAITHFUL INN ENTRANCE.

The center of the building rises eight stories high, surmounted by the lookout that gives a panoramic view of the geyser basin. From half a dozen golden topped flagstaffs float the emblems of various nations. At night, by the aid of a powerful light, one discerns geysers in action, and bears feeding at the edge of the timber. The illumination of Old Faithful Geyser in

OLD FAITHFUL INN.

action is a sight never to be forgotten. The Old Faithful Inn was built at a cost of two hundred thousand dollars and was first opened to the public for the season of 1904.

Old Faithful Geyser.—Less than 1,000 feet east, and in plain sight from the hotel, is located this reliable friend of the tourist. Every sixty-five minutes (with rarely a variation of five minutes) day and night, summer and winter, this wonderful freak of nature gives its exhibition. The position and direction of the sun and wind vary the appearance of this geyser, which is one of the most popular in the Park, because of the remarkable regularity with which its eruptions occur, and the excellent opportunities afforded for observation. Eruptions by moonlight, at sunrise or sunset, in a storm or with clear weather, with their varied effects equally command the attention of the visitor.

OLD FAITHFUL GEYSER AT SUNRISE.

Its eruptions begin with a few spasmodic spurts, during which considerable water is thrown out; these are followed in from five to eight minutes by a column of hot water two feet in diameter, which is projected upwards to a height of 125 to 150 feet, when it remains apparently stationary for about three minutes.

The **Hamilton Curio Store,** formerly called the Klamer Curio Store, has a large variety of supplies and interesting souvenirs, and **The Haynes Picture Shop** carries a complete line of photographs, paintings and art prints of the park.

PARTY OF TOURISTS AND GEYSER HILL.

The **Beehive Geyser** is situated almost in front of Old Faithful Inn on **Geyser Hill** across the river. Its symmetrical cone, shaped like an old-fashioned beehive, is four feet high and three feet across. The Beehive plays out of its nozzle-like opening to the amazing height of two hundred feet.

Eruptions of the Beehive are foretold by the spouting of its indicator, a small, inconspicuous fissure in the formation ten feet north of the cone.

There is undoubtedly some relation between this geyser and the Giantess, a hundred yards higher up on Geyser Hill, because invariably after eruptions of the Giantess, the Beehive plays two, three and some-

Copyright by Haynes, St. Paul.
OLD FAITHFUL GEYSER BATHS.

THE FIRST AUTOMOBILE TO ENTER THE PARK.
August 1st, 1915.

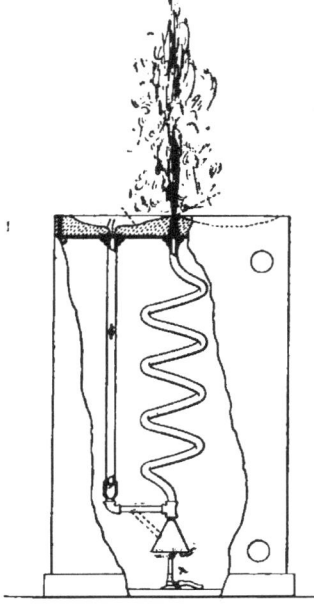

THIS ARTIFICIAL GEYSER
ERUPTS TWO FEET HIGH.

A miniature, mechanical geyser, **"Old Faithful, Jr.,"** is on exhibition at the Haynes Picture Shop, Upper Basin; it accurately demonstrates the Bunsen Theory and is furnished by the publisher of this book. *A real hot water, erupting geyser. Do not fail to see it.*

Free demonstrations daily in the interest of science at 3:30 P. M. and 8:00 P. M.

The playing of the larger geysers is announced by an electric **Siren** at the Haynes Picture Shop.

HAYNES PICTURE SHOP—UPPER BASIN.

Copyright by Haynes, St. Paul.

Copyright by Haynes, St. Paul.

SHAW AND POWELL CAMPING COMPANY BUILDING AND TENTS AT
UPPER BASIN.

times four times, at intervals of eight to twelve hours; and occasionally, but rarely, once *before* the Giantess, but at no other times.

A few feet east of the Beehive cone at the top of the river bank, is the **Cascade Geyser,** now but a quiet spring. Down at the river's edge is the **Sputterer,** which discharges at intervals directly into the river. On the opposite bank is the **Chinaman Geyser,** which was named in memory of the Mongolian who established a laundry here, put in the clothes and soap, and was annihilated, so the story goes, by the violent eruption which ensued. It is a remarkable fact that a bar or two of soap will cause practically any geyser to play within a few minutes. The practice of causing eruptions in this manner became so common a few years ago that the government put a stop to it, as it was feared some of the best geysers would be ruined. It is unlawful to throw any substance into the springs or geyser vents, or to injure, or disturb, in any manner, or to carry off any of the mineral deposits, specimens, natural curiosities, or wonders within the Park.

The **Giantess Geyser** occupies the most prominent position on Geyser Hill. Its displays attain the height frequently of one hundred feet, and are accompanied by shocks and tremors not unlike earthquakes. After the thirty-foot crater of the Giantess is emptied, a steam-period ensues, the entire eruption lasting from twelve to twenty-four hours. During 1911 the records show that the intervals between eruptions varied from four to twelve days; while a few years ago the Giantess played only every three to four weeks. This accurate record disproves, in this case at least, that the geysers are all diminishing in eruptive violence and frequency. It is now pretty generally believed that, while this thermal activity is decreasing as a whole, a century brings only an imperceptible change. The late N. P. Langford, writer and explorer, who visited the Park with the Washburn party in 1870, stated in 1910, while at the Upper Basin, that he saw absolutely no change in Old Faithful Geyser, or any of the others to warrant the assertion that geyser activity is on the decline.

The **Butterfly Spring,** several rods east of the Giantess, is interesting from the fact that its shape and coloring closely resembles a butterfly; this spring is about four feet across and has openings in both "wings."

On the prominence with the Giantess, are two cauldrons, the one having a rim is the **Teakettle,** the other the **Vault;** the latter is a geyser which plays eight feet high twenty-four hours before the Giantess. **Topaz Pool** is at the base of the Giantess mound.

The **Pump,** at the foot of the Giantess mound in the direction of the Sponge Geyser, is a hole eighteen inches across out of which comes a thumping sound at intervals closely resembling an hydraulic ram at work.

Sponge Geyser, a short distance east of the Giantess, is remarkable chiefly on account of the appearance of its cone, a flinty formation, porous and colored like a sponge. The eruptions occur a minute and a quarter apart and are only about four feet high; in reality nothing more than intermittent periods of violent boiling.

Doublet Pool, marked "Dangerous" on the signboard, is a good example of the overhanging crust formation. No doubt in time it will be practically all covered over; although this *sinter* formation, characteristic of the entire Upper Basin, forms very slowly.

UPPER GEYSER BASIN.

Beach Spring, north of the Doublet, is very interesting; it consists of a central opening surrounded by a rather wide, submerged beach, which is symmetrical and practically flat.

The Ear is on the summit of a mound between the Beach and the Lion group. Curiously enough it not only resembles an ear in shape, but the lobe is pierced and the earring is a tiny geyser. It is here that messages are transmitted, so the story goes, to regions below.

The **Lion Geyser,** with the Lioness and two Cubs, occupies a conspicuous mound west of the Giantess and in sight of the hotel. Its eruptions occur usually in series of three, about two and one-half hours apart, following a quiet period of twelve hours. The first eruption of the three is the most spectacular, being about sixty feet high and lasting five minutes.

The **Lioness Geyser** has been observed not to play at all some seasons, while during other seasons eruptions have been noted at intervals of about fifteen days. In 1903 the Lion, Lioness and both Cubs played simul-

HORSEBACK PARTY UNDER DIRECTION OF CHAS. C. MOORE, A NOTED
YELLOWSTONE PARK GUIDE.

taneously one day for a large party of tourists; this remarkable exhibition is attributed (by some) to the completion that year of the famous Old Faithful Inn. The larger Cub plays with the Lioness to a height of thirty feet; the smaller one plays frequently, but only a few feet high.

A path leads from the Lion group past the **Liberty**

Pool to the **Sawmill Geyser,** without crossing the river. The Sawmill gets its name from the peculiar noise accompanying eruptions; the maximum height of this geyser is forty feet, interval three to four hours. Its indicator is a few feet southeast; both the indicator and the Sawmill start together, and very suddenly, throwing water in every direction.

Copyright by Haynes, St. Paul.
GRAND GEYSER—200 FT.

Passing the **Bulger, Tardy,** and **Triplets,** all of minor importance, the Grand group is reached next.

The **Grand Geyser** is one of the finest in the park. It discharges forked columns of water to a height of two hundred feet in a series of ten or twelve distinct eruptions. It is very irregular, playing at intervals varying from one to ten days; its duration is usually

FREDERICK MORRIS, CODY, WYOMING,
A NOTED YELLOWSTONE PARK GUIDE.

from thirty to sixty minutes. The Grand Geyser plays much more frequently in the spring than in the fall, probably as the water supply from the surrounding mountains is greater at that time of the year.

Adjacent to the Grand Geyser crater is the **Turban Geyser,** which plays out of a small fissure next to the main crater of the Turban. When quiet, the larger crater often presents the appearance, in its interior, of a dancing flame, caused by the light playing on the bubbles of gas which constantly arise therefrom. This illusion is so realistic that many of the early explorers really believed that internal fires were visible here. Firehole Lake, at the Lower Basin, also affords a good example of this phenomenon. The Turban plays

twenty-five feet high and at an angle, eruptions lasting an hour or more, and occurring with the Grand
Geyser and at other times.

The fittingly-named **Economic Geyser** is a few rods
north of the Turban; after its eruptions, which occur
at intervals following the Grand Geyser, all the water
expelled flows back into the crater and disappears. The
Economic is only fifteen or twenty feet high, but in
its general form resembles Old Faithful.

Beauty Spring, a large, silent pool, is remarkable for
its beautiful coloring and its highly ornamental margin. **Chromatic Pool**, near Beauty Spring, offers a
good example of colored geyser formation; a rust color
predominates in various shades from yellow to richest
brown, blending into green and delicate pinks. The

OBLONG GEYSER CRATER.

mushroom-like algous growths seen in some of the bordering pools are of interest, even to the casual observer, on account of their peculiar forms and colors, and to the scientist who knows what an important part the algae have in the rate and manner of deposition of silica.

The **Oblong Geyser,** so named on account of the shape of its crater, is on the opposite side of the Firehole River from Chromatic Pool. Its eruptions are of minor importance, the crater being remarkable in that no better example of interior geyser structure is seen in the entire park region. Large globular masses of tan colored geyserite form the rim; the water is a delicate blue color and of such transparency that the two fissures in the bottom of the crater are plainly seen. Preceding eruptions the crater fills to the shore line and boils for fifteen minutes, so the best time to view the crater is immediately after an eruption, when the water level is lowest.

GEOLOGICAL.—A geyser may be defined as a periodically erupting hot spring as its water is not volcanic but simply hot meteoric water; so a geyser is not a volcano ejecting water but a true spring. Were the heat sufficient and the tube long enough all hot springs would erupt.

Sounds like cannonading are heard directly preceding a geyser eruption; this is caused by the collapse of steam bubbles from the hotter region below rising through the cooler strata of water. The surface of the pool, from which the geyser plays, bulges and overflows, and sometimes jets of water are thrown upward preceding activity.

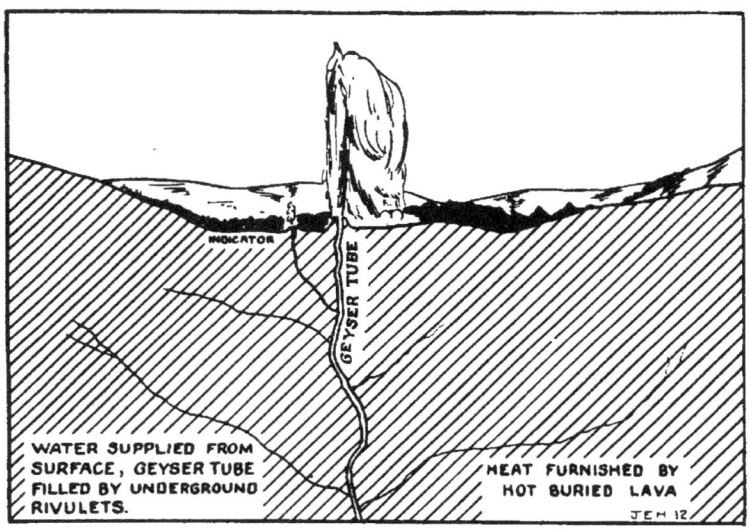

GEOLOGIC PROFILE, TYPICAL GEYSER.

The famous scientist R. W. Bunsen, after making a careful study of geyser action by extensive observation and experiment, advanced the following authoritative explanation:

It is well known that the pressure in water (being due to gravity) increases with the depth; and furthermore, that the boiling point rises with the increase in pressure. The geyser tube which extends deep into the earth is filled with water from the higher tracts of land around; the heat is from the buried masses of lava not yet cool, lava being such a great non-conductor and retainer of heat.

The typical geyser eruption may be divided into five stages, namely, (1) the water remains practically stationary after the tube has filled, and becomes steadily hotter, (2) steam bubbles rising through the cooler strata of water, collapse, producing the characteristic premonitory "cannonading," (3) steam forms below in sufficient quantity to cause the surface to overflow, thus the pressure is lessened in all parts of the tube, and (4) the great burst of steam ensuing, ejects all the water from the tube, (5) the steam follows and while the tube is filling for another eruption, there is no activity other than occasional puffs of steam.

From the Upper Basin to Yellowstone Lake (19 miles) the route is over the summit of the continental divide, near Shoshone Lake, the head waters of Lewis Fork of Snake River (a branch of the Columbia that empties into the Pacific Ocean) ; and in a few miles re-

KEPLER CASCADES.

turns to the Atlantic Slope at Yellowstone Lake, whose waters reach the ocean through the Yellowstone, Missouri and Mississippi rivers.

The road leads up and across the Firehole River and climbs an ascent to

Kepler Cascade, less than two miles distant, whose waters leap from shelf to shelf in a rocky chasm in a series of enchanting falls, aggregating 100 to 150 feet in height, and whose charms are enhanced by the dark background of forest on either hand. The roadway continues up the Firehole about two miles to the third crossing, when it leaves the river, following the course of Spring Creek nearly to the summit of the Divide.

Copyright by Haynes, St. Paul.

TWO-OCEAN ISA LAKE.

UNION GEYSER, SHOSHONE GEYSER BASIN.
The largest geyser in the group; it spouts simultaneously from three craters.

Lone Star Geyser is off the main road and is visited only as a side trip. Its cone, twelve feet high, has a large central opening and numerous adjacent small ones from which water is thrown during eruptions. The cone is the principal attraction of this geyser although the eruptions are at times 75 feet high.

The road leaves the Firehole two miles from Upper Basin and follows the course of Spring Creek nearly to the summit of the Continental Divide.

At a point eight miles from Upper Basin is **Norris Pass** through which a trail leads south to Shoshone Lake. **Craig Pass** is one-half mile further.

Isa Lake is seen next; its waters flow to both the Atlantic and Pacific Oceans from the summit of the Continental Divide. **Two Ocean Pond**, a similar lake, is also on the summit of this range a few miles south of Yellowstone Lake.

The **Continental Divide,** crossed twice between the Upper Basin and Yellowstone Lake, is a great range of mountains extending from Canada to Mexico. It enters the Yellowstone Park near the Western Entrance and extends through the reserve to its southern border forming the water shed between Yellowstone Lake and the headwaters of Snake River.

Shoshone Point affords one of the most commanding views of this ride. It overlooks the country to the south, Shoshone Lake in a beautiful valley, and the Teton Mountains many miles south.

From Shoshone Point the drive is less attractive; however, it again crosses the continental divide at a "pass" so level that it is difficult to know when the summit is really reached.

MACKINAW TROUT CAUGHT IN SHOSHONE LAKE.

Shoshone Lake has an area of about a dozen square miles, with an irregular shore line. Shoshone Geyser Basin, situated on the west shore of the lake, has several large geysers and numerous interesting springs; it is reached by trail from the Lone Star Geyser.

On a clear day from Shoshone Point may be seen the three snow-capped "Sentinels" of the Teton Mountains fifty miles distant, that form a portion of the boundary between the states of Wyoming and Idaho, their dizzy heights, full 14,000 feet, overtopping all other peaks of the Rockies.

Lake View.—A mile from the lunch station at Thumb Bay, one catches the first glimpse of Yellow-

JACKSON LAKE AND TETON MOUNTAINS.

stone Lake. From this point Mr. David E. Folsom, of the Folsom and Cook exploring party in 1869, says: "As we were about departing on our homeward trip we ascended the summit of a neighboring hill and took a final look at Yellowstone Lake. Nestled among the forest-crowned hills which bounded our vision lay this inland sea, its crystal waves dancing and sparkling in the sunlight as if laughing with joy for their wild freedom. It is a scene of transcendent beauty which has been viewed by but few white men, and we felt glad to have looked upon it before its primeval solitude should be broken by the crowds of pleasure seekers which at no distant day will throng its shores."

Thumb Bay Lunch Station (Alt. 7,788 ft.) is pleasantly situated on the shore of Yellowstone Lake facing Thumb Bay and the **Wylie Thumb Camp** is near at hand at the edge of the timber.

YELLOWSTONE LAKE AND SLEEPING GIANT.

Copyright by Haynes, St. Paul.

A SHAW AND POWELL TENT STREET.

The **Yellowstone Lake Boat Company** provides rowboats and launches for both fishing and sight-seeing excursions. The trip to the southeast arm is a most enjoyable one from the fact that the more timid wild animals, including the moose, are often seen along the shore. Rods and fishing tackle may be rented at the various hotels and camps at all points in the park.

Fishing is permitted only with rod and line, and while it is seldom that one catches a fish smaller than eight inches in length all such must be returned to the water with the least possible damage to the fish.

A YELLOWSTONE-WESTERN STAGE COACH PARTY.

The **Shaw & Powell Camp** is a short distance north, pleasantly situated overlooking the lake.

At the Thumb there are several geyser cones, paint pots and springs. The **Paint Pots** are not so large as those at the Lower Basin but they are different.

The **Lake Shore Geyser,** 100 yards from the Hotel, plays at intervals several feet high. The **Fishing Cone** was named by the Expedition of 1870. This cone, with a boiling spring in its centre projects above, and is surrounded by, the cold water of the Lake. This is the famous place where fishermen stand and, after catching trout in the Lake, boil them while still on the hook in the hot spring, (a practice now prohibited by **law.)**

Copyright by Haynes, St. Paul.

Copyright by Haynes, St. Paul.

SINGLE AND TWO-ROOM TENTS.
WYLIE PERMANENT CAMPING COMPANY.

Tourists have an opportunity of taking a steamer or launch from Thumb Bay Lunch Station to the Colonial Hotel at the Lake outlet—a very pleasant ride.

The **Southern Entrance to the Park.**—A wagon road has been constructed south from Yellowstone Lake, passing Lewis Lake and continuing down the valley of Snake River to the southern boundary of the Park and Jackson Hole.

The **Natural Bridge** is passed on the drive around the Lake 3½ miles from the Lake Hotel. It spans a small creek and looks quite symmetrical from the lower side. Its abutments are thirty feet apart, and the arch is sixty feet high.

The **Yellowstone Lake** is the largest at its altitude (7,741 ft.) in the world with the exception of Lake

YELLOWSTONE LAKE BY MOONLIGHT.

Titicaca, Peru. It is twenty miles across and of very
irregular outline. The snow-capped mountains of the
Absaroka Range rise from the water's edge to altitudes of ten or eleven thousand feet.

Several islands dot the surface of this icy sheet of
water, Stevenson and Frank Islands being the largest.

COLONIAL HOTEL, YELLOWSTONE LAKE.

Sheltered as it is, the surface is seldom rough. The
Yellowstone River is at once its principal affluent and
sole outlet, its upper portion draining a considerable
area tributary to the lake on the southeast, and the vast
body of water thus accumulated in this natural mountain reservoir serves not only to furnish a never-failing
supply for one of the grandest of the Missouri's tribu-

taries, but supplies the means of successful irrigation
of the entire lower Yellowstone Valley.

Sleeping Giant.—In the mountain range on the
east side of the lake can be seen the "Sleeping Giant."
It is formed of the peaks of Saddle Mountain in connection with a mountain range several miles this side.

FISHING BOATS AT LAKE OUTLET.

Fishing Grounds.—In the river at the lake outlet
about a mile from the hotel are the fishing grounds.
During the trout season (July to September), no better fishing can be found. The fish average about one
and one-half pounds each and are of the *salmo mykiss*
variety. A catch of 20, three or four hours before
sundown, is not infrequent.

Hotel at the Outlet.—This spacious and elegantly appointed hotel tends greatly toward making Yellowstone Lake the resort par excellence of the Park. Here everything is so arranged that guests can spend the entire season, if they so desire, making short, easy trips to all points of the great reserve.

To visit any or all of the points circumjacent to this grand mountain lake, vehicles of all kinds, saddle and pack animals, guides, rowboats, launches and steamers, are ever at command. The **Wylie Lake Camp** is pleasantly situated near the lake outlet on the west of the road.

The Eastern or Cody Entrance to Yellowstone National Park is considered by many travelers the most

LOCH LEVEN AND RAINBOW TROUT.

(Photo courtesy The Cody Club.)
SHOSHONE DAM.

picturesque. It is 57 miles from Cody, Wyoming, on the Chicago, Burlington & Quincy Railroad to the eastern boundary and 28 miles further to the Yellowstone Lake on the main circle road.

This distance is covered in one day by the Cody-Sylvan Pass Motor Company, organized in 1916 to transport passengers from Cody to Yellowstone Lake and return. At Yellowstone Lake the passengers transfer to the vehicles of the various companies operating over the regular route.

Near and in Cody there are summer resorts where the rates are moderate and the facilities the best. Private camping and horseback parties are made up at this place for a tour of the park.

(Photo courtesy The Burlington Route.)

THE CODY ROAD.

(*Photo by Lucier.*)

SYLVAN LAKE—CODY ROAD.

The road leads up Shoshone River through a valley and Shoshone Canyon, and up Middle Creek to Sylvan Pass. One mile beyond the soldier station at the eastern entrance is a good camping place.

Through the Shoshone Canyon the road has been chiseled in and through the rock, forming hanging roadways and long tunnels, and in some places it is so high above the river that the sound of the rushing water cannot be heard.

Eight miles from the eastern boundary is Sylvan Pass at an altitude of 8,650 feet; and a mile beyond is Sylvan Lake, altitude 8,350 feet. Twenty miles from the eastern boundary is Turbid Lake, which is remarkable for the many hot springs and steam vents along its shores and under the water.

Yellowstone Lake to Falls and Canyon.—From the lake to the Grand Canyon, a distance of seventeen miles, the road follows the valley of the Yellowstone, and passes through Hayden Valley.

Mud Volcano is 7½ miles from the Lake Hotel, a few rods west of the road on the mountain side; its funnel-shaped crater is 30 feet deep and is partly filled with a lead-colored mass of mud in violent agitation, producing an effect at once repulsive and fascinating. In 1898 most violent eruptions occurred, and the surrounding trees were plastered with mud.

Green Gable Spring a few rods north of the Mud Volcano, is a beautiful overflowing hot pool beneath a

GRAND CANYON BRIDGE.

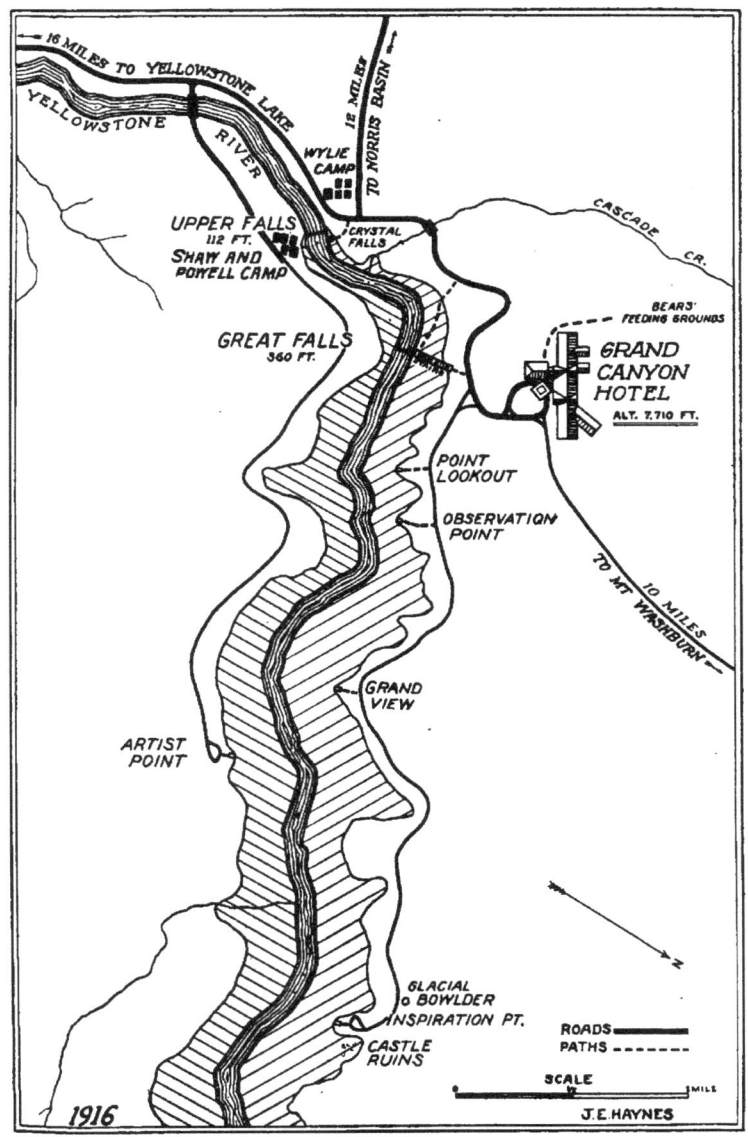

GRAND CANYON OF THE YELLOWSTONE.

Copyright by Haynes, St. Paul.

TOURISTS AT SHAW AND POWELL CANYON CAMP.

natural rock gable colored a rich green. Its peacefulness pleasantly contrasts with the violence of the Mud Volcano.

The Grand Canyon Bridge is the largest bridge of its kind in the world. The road across this bridge reaches Artist Point, from which one may enjoy by far the best view of the Falls and Canyon.

The **Upper Falls** has a perpendicular drop of 109 feet, and the water, striking the shelving rock formation at the bottom of the abyss, shoots out in rocket-like jets. Above the falls a jutting point of rock affords an excellent view of the rapids and the foaming waters as they rush over the precipice. A footpath leads to the bottom of the Upper Falls, where very fine brook trout fishing may be enjoyed. Midway between this

GREAT FALLS OF THE YELLOWSTONE—308 FEET.

point and the Lower Falls, Cascade Creek enters the river.

Crystal Falls is below the bridge which spans the creek. The aggregate fall, including the cascades above, is about 130 feet, and a ladder to Grotto Pool allows an inspection of them.

The **Wylie Canyon Camp,** a night station, is located near Canyon Junction; and the **Shaw & Powell Canyon Camp** is across the river near the Upper Falls.

The **Great Falls** of the Yellowstone, 308 feet in height, is a quarter of a mile below the Upper Falls. Not far from the Canyon Hotel is a recently-built stairway descending to the very brink of the Great Falls. This view overlooking the awful plunge of

UPPER FALLS OF THE YELLOWSTONE—109 FEET.

seething waters twice as high as Niagara, is grand almost beyond expression.

Gazing down the canyon from the brink of the Falls one sees Point Lookout 1,200 feet above the river. Almost directly opposite, on the right hand side of the canyon, is Artist Point.

The **New Grand Canyon Hotel,** situated at the Great Falls and Grand Canyon, was first opened to the public June 15th, 1911. It cost over three-quarters of a million dollars, and has accommodations for over five hundred guests.

Spending the day at this hotel is a pleasure. A cozy foyer, extensive lounge and capacious dining room,

Copyright by Haynes, St. Paul.

GRAND CANYON FROM THE BRINK.

GRAND CANYON HOTEL.

each elegantly furnished and of novel architecture, are exceptionally attractive. The service, too, is above criticism; every convenience for the traveler is supplied.

It is remarkable that so many miles from any railroad hotels can be so well supplied with equipment and food that they rival the best hostelries in the large cities. The electric lights are artistically arranged in ornamental hanging lanterns. The building has telephone and elevator service throughout.

Adjoining the main part of the building

GRAND CANYON HOTEL OFFICE.

is the lounge, where concerts and dances are held. Adjoining it are tea and buffet rooms.

Mr. Robt. C. Reamer, architect, designed and built this hotel against great odds. Three hotels belonging to the Yellowstone Park Hotel Company stand to his credit—the Old Faithful Inn, the largest log structure in the world; the Lake Hotel, which he remodeled and enlarged, and this, the Grand Canyon Hotel, which is a masterpiece.

Point Lookout.—The driveway follows, as nearly as practicable, the very edge of the canyon from the falls to Inspiration Point, about three miles. Point Lookout is about half a mile below the falls, and commands altogether the best combined view of the Great Falls and Grand Canyon. It is fully 1,200 feet above the

GREAT FALLS OF THE YELLOWSTONE—308 FEET.

river and nearer the hotel than any of the several points of observation. **Red Rock** under Point Lookout, to which a perfectly safe trail leads down the ravine near the point, affords the best view of the falls possible for tourists to obtain.

Grand View.—There are many projections between Lookout and Inspiration Points, from which glimpses of the canyon may be had. Grand View is about midway between Point Lookout and Inspiration Point, nearly opposite **Artist Point** on the opposite side of the canyon. It affords an excellent view of the canyon and of the rugged cliffs about Inspiration Point.

Inspiration Point is considered the best place from which to see and appreciate the immensity of the canyon; it is two miles from Point Lookout and over 1,000 feet above the river.

Copyright by Haynes, St. Paul.

ROCK SPIRES NEAR TOWER FALLS.

Copyright by Haynes, St. Paul.

YELLOWSTONE CANYON NEAR TOWER FALLS.

Copyright by Haynes, St. Paul.

OVERHANGING CLIFF ON TOWER FALLS ROADWAY.

Glacial Bowlder is passed on the drive from the hotel to Inspiration Point on the north side of the canyon.

This huge block bespeaks the great transporting power of the glaciers; it is alone among mountains and canyons of a finer rock texture and was brought from a point many miles distant from its present resting place.

The interestingly graphic and faithful pen picture of the **Grand Canyon and Great Falls of the Yellowstone,** by the Rev. Dr. Wayland Hoyt, follows:

"Look yonder! Those are the Lower Falls of the Yellowstone. They are not the grandest in the world, but there are none more beautiful. There is not the breadth and dash of Niagara, nor is there the enormous depth of leap of some of the waterfalls of the Yosemite. But here is majesty of its own kind, and beauty, too. On either side are vast pinnacles of sculptured rock. There, where the rock opens for the river, its waters are compressed from a width of 200 feet between the Upper and Lower Falls, to less than 100 feet when it takes the plunge. The shelf of rock over which it leaps is absolutely level. The water seems to wait a moment on its verge; then it passes, with a single bound, 308 feet into the gorge below. It is a sheer, unbroken compact, shining mass of silver foam. But your eyes are all the while distracted from the fall itself, great and beautiful as it is, to its marvelous setting; to the surprising, overmastering canyon into which the river leaps, and through which it flows, dwindling to but a foamy ribbon there in its appalling depths. As you

POINT LOOKOUT AND GREAT FALLS.

cling here to this jutting rock, the falls are already many hundred feet below you. The falls unroll their whiteness down amid the canyon glooms. * * * These rocky sides are almost perpendicular; indeed, in many places the boiling springs have gouged them out so as to leave overhanging cliffs and tables at the top. Take a stone and throw it over; you have to wait long before you hear it strike. Nothing more awful have I ever seen than the yawning of that chasm. And the stillness, solemn as midnight, profound as death. The water dashing there, as in a kind of agony, against these you cannot hear. The mighty distance lays the finger of silence on its white lips. You are oppressed with a sense of danger. It is as though the vastness would soon force you from the rock to which you cling. The silence, the sheer depth, the gloom burden you. It is a relief to feel the firm earth beneath your feet again, as you carefully crawl back from your perching place.

"But this is not all, nor is the half yet told. As soon as you can stand it, go out on that jutting rock again and mark the sculpturing of God upon those vast and solemn walls. By dash of wind and wave, by forces of the frost, by file of snow plunge and glacier and mountain torrents, by the hot breath of boiling springs, those walls have been cut into the most various and surprising shapes. I have seen the 'middle age' castles along the Rhine; there those castles are reproduced exactly. I have seen the soaring summit of the great cathedral spires in the country beyond the sea; there they stand in prototype, only loftier and more sublime.

"And then, of course, and almost beyond all else, you are fascinated by the magnificence and utter opulence of color. Those are not simple gray and hoary depths, and reaches and domes and pinnacles of sullen rock. The whole gorge flames. It is as though rainbows had fallen out of the sky and hung themselves there like glorious banners. The underlying color is the clearest yellow; this flushes onward into orange. Down at the base the deepest mosses unroll their draperies of the most vivid green; browns, sweet and soft, do their blending; white rocks stand spectral; turrets of rock shoot up as crimson as though they were drenched through with blood. It is a wilderness of color. It is impossible that even the pencil of an artist can tell it.

What you would call, accustomed to the softer tints of nature, a great exaggeration, would be the utmost tameness compared with the reality. It is as if the most glorious sunset you ever saw had been caught and held upon that resplendent, awful gorge.

"Through nearly all the hours of that afternoon until the sunset shadows came, and afterwards amid the moonbeams, I waited there, clinging to that rock, jut-

Copyright by Haynes, St. Paul.

ROADWAY ASCENDING MOUNT WASHBURN.

ting out into that overpowering, gorgeous chasm. I was appalled and fascinated, afraid, and yet compelled to cling there. It was an epoch in my life."

Grand Canyon to Norris Geyser Basin.—The twelve-mile drive between the Canyon and Norris is through an undulating pine forest the greater part of the distance. It passes over a "divide" separating the Yellowstone and Missouri Rivers at an altitude of more than 8,000 feet. Between the five and six-mile posts, on the south side of the road are the **Wedded Trees.** The **Virginia Cascades** are about three miles from Norris, being quite unlike the many falls and cascades seen throughout the park.*

MOUNT WASHBURN—10,388 FT.

A **Side Trip** from the **Canyon** to **Mammoth Hot Springs** via Dunraven Pass and Tower Falls may be made when conditions are favorable. It is about ten miles to Tower Falls from Mt. Washburn. Near Tower Falls is the Y-W Stage Company's **Relay Station;** the Soldier Station and the **Wylie Camp** are

*See page 130 for continuation, Norris to Yellowstone. See page 113 for tour from Mammoth to Norris.

about a mile and a half from the falls, the latter
equipped to accommodate a limited number of visitors
over night if desired.

Tower Falls is 110 feet high; near it are the tall
rock spires unlike any other lava formation in the
park. Across the Grand Canyon from the mouth of
Tower Creek the canyon wall is capped with a thick
layer of lava which is distinctive of this portion of the
park.

TOWER FALLS—110 FT.

Copyright by Haynes, St. Paul.

PETRIFIED TREES.

The **Petrified Trees** are situated one-half mile south of the main roadway, about eighteen miles from Mammoth Hot Springs; they are reached by a side road, and consist of two large stumps standing in their natural positions on the hillside.

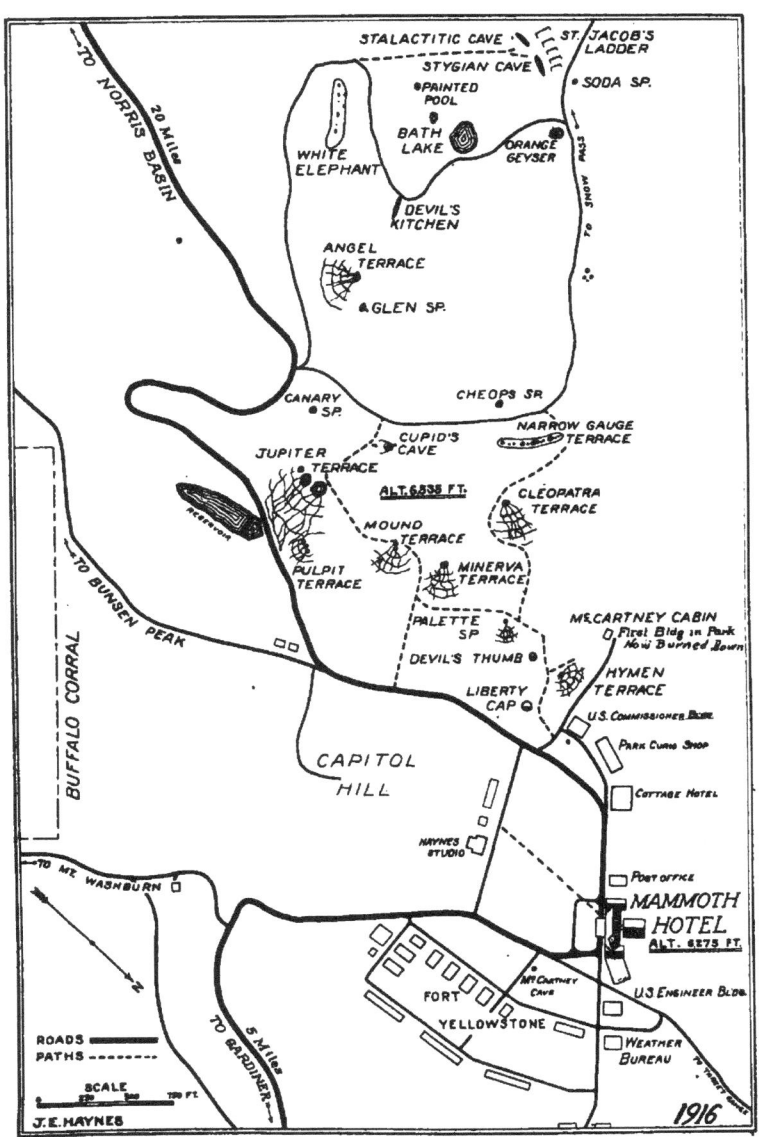

MAMMOTH HOT SPRINGS AND FORT YELLOWSTONE.

(Photo courtesy White Automobile Co.)
NORTHERN ENTRANCE ARCH AND FIRST AUTOMOBILE IN PARK—
AUGUST 1, 1915.

TOUR OF THE PARK
FROM THE NORTHERN ENTRANCE

Northern Entrance Arch.—In 1903 the government built an imposing stone arch at Gardiner, dedicated by President Roosevelt, which bears the following inscriptions: "Yellowstone National Park," "Created by Act of Congress March 1, 1872," "For the Benefit and Enjoyment of the People." The log depot of the Northern Pacific Railway in Gardiner, the terminus of the park branch, is in keeping with its mountain surroundings.

From Gardiner to Mammoth Hot Springs (5 miles) the road leads through Gardiner Canyon past the picturesque **Eagle Nest Rock,** and the **Boiling River.** On this drive an ascent of nearly a thousand feet is made.

Fort Yellowstone, at Mammoth Hot Springs, is the administrative headquarters of the park. It recently has been greatly enlarged by the addition of a number of large stone buildings and stables made out of lava rock quarried in the vicinity. Small posts are maintained at various places in the park for the cavalrymen who patrol the roads and police the park in general.

Copyright by Haynes, St. Paul.

LIBERTY CAP AND MAMMOTH HOTEL.

Mammoth Hotel (Alt. 6,275 feet) is on the same plateau, near Fort Yellowstone, with the U. S. Commissioner's building, the U. S. Engineer Office and the Weather Bureau.

Mammoth Hot Springs (Alt. 6,275 to 6,575 feet).— The hot springs and terraces occupy several acres to the south of the plateau on the slope of Terrace Mountain. To visit all the prominent springs and formations requires fully two hours; from the road to Golden Gate, however, an excellent view of Jupiter Terrace, the largest of the group, is obtained.

To thoroughly inspect these wonderful springs one must do considerable walking. The "Formation Party" accompanied by a competent guide starts from

Mammoth Hotel early in the afternoon for a two-hour stroll among these wonders.

One characteristic of this great lime deposit (travertine) is the absence of color where dry. The beautiful colorings which have made the terraces famous appear only where the water flows; when the overflow from any spring changes its course the algae, which produce the color, disappear from the abandoned runway and soon the new course is brilliantly colored.

Hymen Terrace, one of the most beautifully colored spots in the Park, is located near the hotel not far to the right of Liberty Cap. This new addition to the number of district terraces at Mammoth, though not

FORT YELLOWSTONE, THE ADMINISTRATIVE HEADQUARTERS
OF THE PARK.

so large as Jupiter, is easily the gem of the collection because of its exquisite coloring. The veil of steam softens and blends the vivid colorings, while innumerable water-glazed knobs reflect the sunlight like a thousand mirrors. Hymen Terrace is growing fast; in fact, it is gravely feared that the openings may become choked from the abundance of depositing lime. If this should happen, it would be a matter of but a few days until the coloring would have disappeared, leaving the travertine rock bare and exposed to the destructive forces of the elements.

McCartney's Cabin, now burned down, in the gulch near Hymen Terrace was of interest historically. In 1877 it was the scene of encounters with the Nez Percé Indians, and was the first building in the park.

Liberty Cap, an extinct hot spring cone, standing at the foot of Terrace Mountain, near the road, is fifty-two feet high and twenty feet in diameter at its base. It is formed of over-lapping layers of deposit, evidently having been built up by the overflow of water through the orifice in the top.

Devil's Thumb, a smaller cone of a similar history and structure, is partially imbedded in the hillside some 200 feet west of Liberty Cap. The path leading to the Formation past the Devil's Thumb is generally taken when returning from the Formation, the one for the ascent branching off the main road a short distance south of Liberty Cap.

Minerva Terrace is a mass of Travertine deposit forty feet in height and covers an area of nearly three-fourths of an acre. The hot spring on the summit is some twenty feet in diameter and at the edge has a temperature of 154 degrees Fahrenheit. The variation of the overflow and the intermittent character of the spring, make it impossible to predict a season in advance which side of the terrace will be the active or whether it will be active at all.

HOT SPRING TERRACES.

Cleopatra Terrace, a short distance above Hymen Terrace, is a good example of the growing deposit. The predominating color here is dark orange.

Jupiter Terrace, the largest of the entire group, extends some 2,000 feet along the edge of the high mound of brilliantly colored deposit south of Minerva Terrace. A climb of about 100 feet up a steep trail, is necessary to reach the summit. The two large springs of boiling water fully 100 feet in diameter, supply the main terrace, as well as the beautiful **Pulpit Terrace** beneath. Articles of iron, glass or any hard substance placed where the water can run over them, are soon coated with a crystal-white deposit of calcium carbonate.

Jupiter Terrace presents the most delicate coloring from the lightest cream through the predominating orange to deep shades of yellow.

Cupid's Cave, situated a few rods west of the great pool on Jupiter Terrace, formerly had an opening that one could enter; but a mass of exquisitely colored stalactite formation has now completely filled it.

Narrow Gauge Terrace, a fissure ridge 300 feet long, is filled when active with miniature geysers and springs which deposit the most brilliant coloring.

Orange Geyser, on the terrace above Narrow Gauge, is greatly admired by all visitors. It consists of an oblong mound of deposit some twenty feet high and about thirty feet in diameter. The active little geyser on its summit and the brilliant coloring are its chief attractions.

Bath Lake, a few hundred feet south of Orange Geyser, is separated from it by a timber-covered ridge of ancient deposit that nearly surrounds the lake. There is no visible outlet to Bath Lake, and the uniform temperature of the water at all seasons of the year is one of the mysteries of this region.

Devil's Kitchen, the crater of an extinct hot spring, can be entered with safety through a small opening

WILD DEER NEAR MAMMOTH HOT SPRINGS.

some six or eight feet in diameter. The hot, damp atmosphere of the interior produces a queer sensation, and the desire to seek fresh air at once comes over the visitor. When the Devil's Kitchen was first explored in 1881 numerous bones of wild animals were found in the cave and it was alive with flying bats.

Stygian Cave.—The poisonous gases from vents in the floor of the Stygian Cave have claimed the lives of many birds and small animals. It is situated a few rods from the Devil's Kitchen and is shaped like a large open fireplace without a chimney; in this chamber collect the gases which suffocate the birds and animals seeking shelter there.

Angel Terrace, one of the most beautifully colored terraces at Mammoth, may be seen through the trees from the Golden Gate Road. It is about three hundred yards south of Orange Geyser.

GEOLOGICAL.—The Yellowstone Park is geologically young but so old that the slow erosive power of running water has carved furrows a thousand feet or more into its solid rock.

The mountains are igneous; and all through the Park are evidences of violent volcanic eruptions as shown by extensive lava beds. Amygdaloid cliffs and great gnarled masses are common; there are obsidian cliffs, great geometrical blocks, petrifactions and geodes, besides the print of leaves in rock where forests have fallen prey to the flowing molten lava.

Some sedimentary deposits are also found here near the northern boundary, in the form of limestone beds, clays and shales. There were glacial invasions from the north too, which have left hills of sand and gravel, and isolated bowlders at various points.

But the most wonderful deposit in the region is this **Formation at Mammoth Hot Springs.**

It is composed of pure calcium carbonate, dissolved from the limestone beds below and brought to the surface by the hot springs. It is many acres in extent —of unknown depth—and is the result of periods of successive deposition and decay extending over a great length of time. The deposit is building where overflowed by water and crumbling to a chalky powder where dry.

The water is heated by great masses of rock which have not yet cooled below the zone of percolating water. Such conditions are also seen today in New Zealand and Iceland.

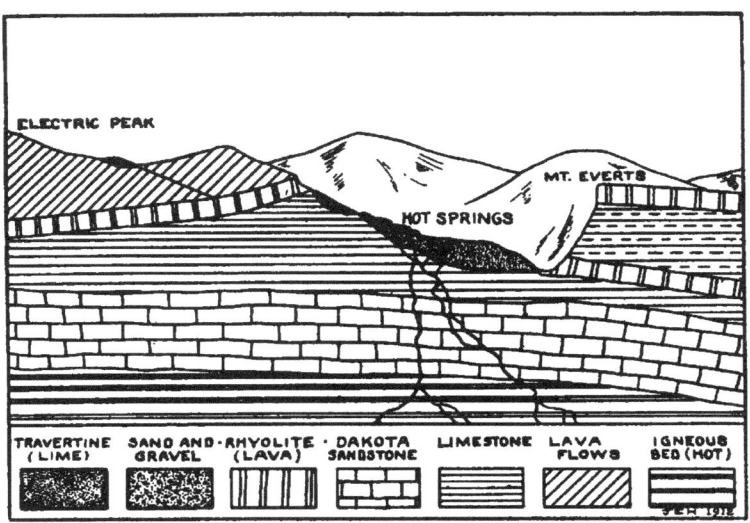

GEOLOGIC PROFILE, MAMMOTH HOT SPRINGS.

Four factors are held responsible for the practically complete precipitation of the lime carried by the water to the surface; namely, (1) the "eating" process of the algous growth which thrives in the hot water, (2) the giving off of carbonic acid to the air, (3) the cooling of the water and (4) evaporation.

The chief attraction of this great deposit is its beautiful coloring; harmonizing shades of yellow and brown with occasional streaks of dark green and red characterize the formation where the hottest water flows. The predominating rust color is found in the tepid water farther from the mouths of the hot springs. It is noticeable that the abandoned portions of the deposit are a glaring chalk-white, also that the colorings are found only on the active terraces; further-more, the color disappears in winter when the water is cooled to the mouths of the boiling springs. Mineral coloring is more stable than that; such coloring re-mains on rock wet or dry, and in a great range of temperatures. It is the algae that color these terraces more beautifully than could natural mineral coloring or the hand of man; the algous growth—a low form of plant life—cleaves closely to the rock in a velvet-like covering and requires hot or tepid water in which to live.

Nor are the pool colorings due to minerals; the United States Geological Survey states authoritatively that these colors are due to the reflection and refrac-tion of the light rays, influenced by the nature and color of the pool linings and their surroundings.

From Mammoth Hot Springs to Norris Basin (20 miles) many interesting places are passed, the great limestone Hoodoos, Golden Gate, Apollinaris Spring, Obsidian Cliff and Roaring Mountain; a very pleas-ant and diversified drive.

Silver Gate and Hoodoos.—The driveway from Mammoth to Golden Gate ascends the mountain by such easy grades and graceful curves that one does not realize that a thousand feet elevation is gained in less than three miles. This road passes through the limestone Hoodoos, a wild region heretofore inacces-sible. Many theories are advanced as to the origin of the "Hoodoos." The most plausible is that the im-mense quantity of deposit or formation seen lower

SILVER GATE AND BUNSEN PEAK.

down the valley, even as far as Gardiner River, two miles distant, was carried there in solution by the hot waters of Mammoth Springs, thus leaving honey-combed caves beneath; the present Hoodoo region was formed by the surface caving in, filling the cavern below with huge masses of fractured rock. This condition is seen over a total area of about one square mile. In the midst of the "Hoodoos" the road makes an abrupt turn, passing between great blocks of limestone that rise abruptly fully seventy-five feet, to which is applied the very appropriate name, **"Silver Gate."**

Golden Gate.—Four miles from Mammoth Springs is one of the most picturesque drives in the Park;

a rugged pass between the base of the lofty eleva-
tions of Bunsen Peak and the southern extremity of
Terrace Mountain (through which flows the west
branch of Gardiner River). The sides of these rocky
walls, which rise 200 to 300 feet above the roadway,
are covered with a yellow moss, suggesting the
name the pass now bears. The pillar at the east
entrance, some twelve feet high, was originally a part
of the canyon wall. The construction of this roadway
and viaduct, scarce a mile in length, was the most
expensive and difficult piece of road building yet encountered by the government engineers.

Rustic Falls, occupying a conspicuous position at
the west end of Golden Gate Canyon, adds a charm

GOLDEN GATE CANYON AND VIADUCT.

to this beautiful spot; in the early part of the season the falls is especially fine. The stream, Glen Creek, is fed by mountain snows and springs, along the base of the hills, a mile or so away; at the falls it leaps some sixty feet over a series of shallow basins worn into the dark, moss-covered ledge, and disappears underneath an accumulation of rock deposited in the canyon when the roadway was constructed.

Swan Lake Basin.—A pleasant surprise awaits the visitor immediately beyond Golden Gate where the road comes suddenly into a broad mountain prairie hemmed in by snow-clad peaks. The magnificent Gallatin range rising abruptly from the foothills, com-

SWAN LAKE CAMP.
WYLIE PERMANENT CAMPING CO.

posed of Bell Peak, Quadrant Mountain, and Mount Holmes (Alt. 10,578 feet), are conspicuous in the foreground. About eight miles to the north is **Electric Peak** (Alt. 11,155 feet), *the highest mountain in the Park,* which contains a large amount of magnetic ore and attracts lightning during storms.

The **Wylie Swan Lake Camp** is conveniently situated at the south end of Swan Lake Basin.

The **Shaw & Powell Willow Park Camp** is on Willow Creek a short distance beyond.

Apollinaris Spring is on the east side of the road

WILLOW PARK CAMP.
SHAW AND POWELL CAMPING COMPANY.

OBSIDIAN CLIFF—BEAVER LAKE.

near the ten-mile post—a delicious spring of natural Apollinaris water, as refreshing as the genuine article of commerce.

Obsidian Cliff.—This bold escarpment of volcanic glass is twelve miles south of Mammoth Hot Springs. The roadway passes along its base for 1,000 feet between it and Beaver Lake. The vertical columns of pentagonal-shaped blocks of obsidian, rising some 250 feet above the road, present a glistening, mirror-like effect when illumined by the sun's rays. The greater part of this mineral glass is jet black and quite opaque, with streaks of red and yellow. The construction of the roadway was accomplished in a novel manner and with considerable difficulty; blasting powder being ineffectual, great fires were built around the huge blocks of glass, which, when heated, were suddenly cooled by dashing water upon them, thus shattering the blocks into small fragments. This process made possible the construction of this really wonderful roadway, probably the only piece of glass road in the world. There being no other exposed ridge of obsidian in the Rocky Mountains, and this material being more desirable than flint for the manufacture of arrow heads, it was once a famous resort for all tribes of Indians, who congregated here in great numbers. Obsidian Cliff was "neutral ground" to all the Rocky Mountain Indians, and undoubtedly as sacred to the various hostile tribes as the far-famed Pipestone country of Minnesota. Chips of obsidian and specimens of partly finished arrow heads are found throughout the Park, generally at places occupied by the Indians as summer camps.

Beaver Lake.—The roadway continues along the east side of Beaver Lake, which is about one mile long and a quarter of a mile wide. Several beaver dams are constructed across the lake, forming a series of artificial obstructions, each having a fall of from two to four feet. A beaver house, still inhabited, is located near the west shore of the lake. Since the rigid enforcement of the Park regulations regarding the killing of game, Beaver Lake is becoming alive with numerous water fowl, the passing carriages not seeming to alarm them. The reflection of the pine-clad hills among the dense growth of pond lilies which line its shores, adds to the beauties of this lake.

The drive from Obsidian Cliff to Norris Basin is over a ridge which separates the headwaters of the

Yellowstone and Missouri Rivers, the ascent of which is so gradual it is impossible to know when the "divide" is passed.

About 4½ miles from Norris, **Roaring Mountain** is seen steaming from countless openings in its furrowed sides. Its ashen color and the muffled sound of escaping steam, less audible now than in the past, make this sight one to be long remembered. Near the roadside at the base of the mountain are greenish, milky pools fed by rivulets of sulphur water from the springs.

Twin Lakes, about four miles from Norris, are remarkable for their beautiful colors. Although situated adjacent to each other they are of decidedly different hues.

The next object of interest is the **Frying Pan,** a basin fifteen feet across, completely filled with little hot springs, or steam vents, which are constantly in a state of violent agitation.

Norris Geyser Basin.—This remarkable geyser region was formerly called "Gibbon Geyser Basin," but on account of the extensive work of exploration done by Colonel P. W. Norris while he was Superintendent of the Park (1877 to 1882), it has since been known as Norris Geyser Basin.

The chief attractions here are the great steam vents, large boiling pools and several geysers, notably the Constant, Whirligig, Mud, Valentine, New Crater, Minute Man, Echinus, Monarch, and Fearless.

Norris Lunch Station (Alt. 7,410 feet) is well situated on a prominence at the north end of the basin

GEYSER TABLE.
NORRIS GEYSER BASIN.

Corrected by observations made during the past season.

Geysers at NORRIS BASIN	Maximum Height	Duration	Intervals of Eruption
Constant..............	20 ft.	10 sec.	30—60 sec.
Echinus..............	30 ft.	irregular	45 min.
Fearless..............	25 ft.	15 min.	3 hrs.
Minute Man..........	15 ft.	1 to 3 min.	1 to 3 min.
Monarch..............	50 ft.	6 min.	25 to 60 min.
Mud.................	20 to 60 ft	1 to 2 min.	New; irregular.
New Crater..........	20 ft.	1 min.	3 min.
Valentine............	100 ft.	40 min.	22 to 30 hrs.
Whirligig............	10 to 15 ft	10 sec.	irregular.

and overlooks the main part. The hotel guide makes two trips over the formation daily.

Congress Pool.—The first sight that attracts the visitor is this immense boiling spring close to the road on the left as he enters the basin. For many years it

Copyright by Haynes, St. Paul.

NORRIS GEYSER BASIN.

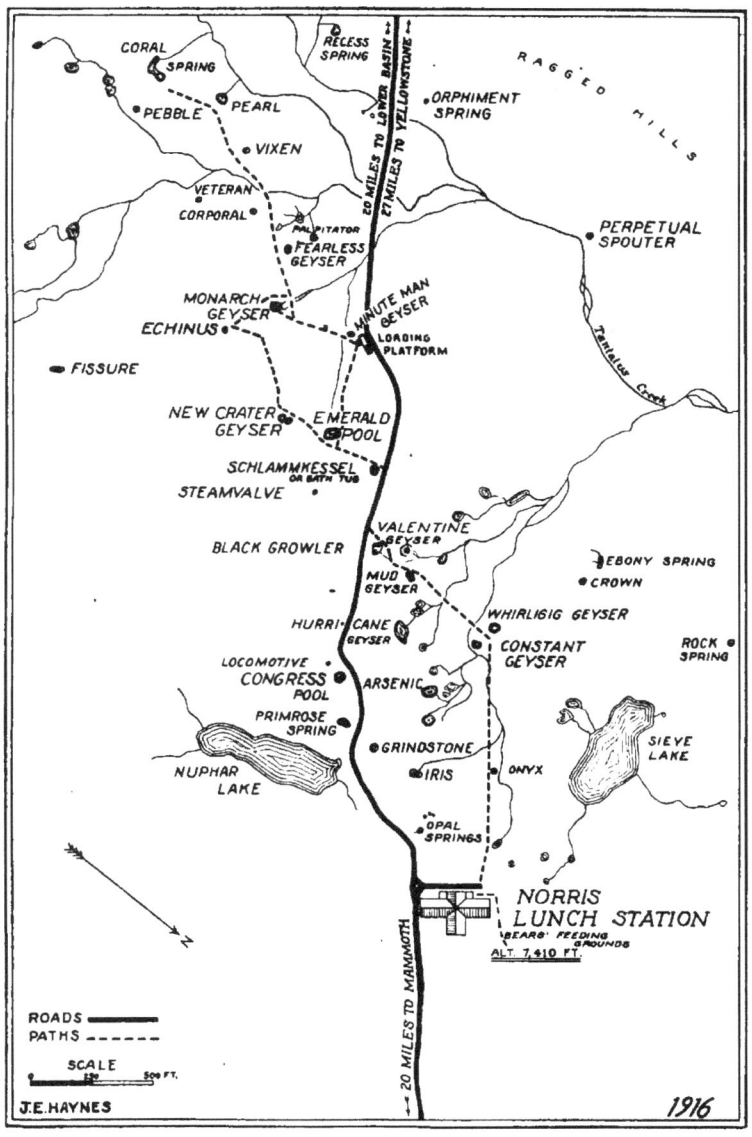

NORRIS GEYSER BASIN.

was only an opening in the rocks from which a great quantity of steam was constantly escaping; the roaring of which could be heard for miles. During the winter of 1893 the "Steam Vent" ceased and the Congress appeared. The first eruptions were of great force and completely blocked the road with masses of earth and formation.

The usual trip over the formation starts from the hotel. To the left of the board walk are **Opal Springs,** the **Iris Pool** and the **Grindstone,** all hot, boiling pools. Near the walk on the right is the **Onyx,** a small basin, and where the walk turns are two geysers, the **Con**stant and Whirligig.

The **Constant Geyser** has a basin twenty-four feet across, out of which displays take place with marked regularity every thirty seconds; a very pretty geyser. A few feet to the south is a similar basin, the crater of the **Whirligig,** which plays quite like the Constant but not so frequently.

The **Mud Geyser** is passed on the way to the Valentine and Black Growler. Some seasons this geyser erupts with great violence, displays frequently occurring about sixty feet high.

The **Valentine Geyser** plays usually every seven and one-half hours, its displays being unequaled by any other geyser in Norris Basin, height, 100 feet, duration, 40 minutes.

Black Growler Steam Vent attracts much attention; it roars constantly and emits great volumes of steam. The deposit around the crater is quite black in places. The vent a few yards north of the Black Growler is known as the **Hurricane**; it is quite similar but not so violent as the former.

Situated east of the roadway is the **Schlammkessel,** frequently referred to as the **Bath Tub.** It has a well-formed basin, and while it does not erupt, it is in constant agitation.

CONSTANT GEYSER—NORRIS BASIN.

Emerald Pool is seen next; a large quiescent lake of boiling hot water with a greenish tinge, situated south of the Bath Tub.

New Crater Geyser.—This geyser is about 500 feet southeast of Emerald Pool, surrounded by huge blocks of dark yellow rock. It came into prominence during the fall of 1891, when quite a commotion, not unlike

NEW CRATER GEYSER—NORRIS BASIN.

an earthquake, was observed. When it burst forth a great volume of water was forced out, flooding the ravine leading to the valley below. Since then it has settled down to ordinary eruptions, about every three minutes. The rock-covered crater prevents the discharge from attaining any great height.

Monarch Geyser, the king of geysers in Norris Basin, is situated at the base of the hill, nearly surrounded by a bluff of brilliantly colored rocks, upon the level of the plateau about 1,000 feet east of the roadway. The crater consists of two oblong openings, the larger of which is twenty feet long and three feet vide. Eruptions of the Monarch occur without warning and consist of a series of explosions, frequently more than a dozen, in which columns of water are thrown 100 feet high. The intervals of eruptions are ordinarily about six hours.

Fearless Geyser, situated 500 feet south of the Minute Man Geyser, throws jets of water in every direction during eruptions. Norris is the newest geyser basin in the Park and probably the one most rapidly changing. One cannot be sure a season in advance whether any one of its geysers will be doubly active the coming summer or become entirely inactive.

The **Minute Man Geyser** is interesting on account of its regularity, and the fact that most of the water thrown out flows back into the crater after the eruption. Its crater is small and appears to have been originally only a fissure in the rock.

The drive to the Fountain Hotel (20 miles) is through Gibbon Canyon past Gibbon Falls and Beryl Spring near the junction of the Gibbon and Firehole rivers, where the road joins that from Yellowstone— the western entrance to Yellowstone Park.

Three miles from Norris Basin the road enters **Elk Park,** a beautiful valley surrounded by heavily-timbered hills.

Chocolate Spring, an unique hot spring has built a cone of rich chocolate color across the river from the road. Further on are the Gibbon Meadows.

At the northern entrance to Gibbon Canyon on the opposite side of the river from the road is Mount Schurz, on the summit of which is the **Monument Geyser Basin** a thousand feet above. Unless one is inclined to scientific observation, a climb up the steep trail to this basin is hardly justified. A dozen or so crumbling geyser cones, some steaming and rumbling, others apparently extinct, constitute its total attractiveness.

Gibbon Canyon.—The roadway enters Gibbon Canyon on the east side of the river, which it follows, as nearly as practicable, for three or four miles, shadowed by precipitous cliffs in places two thousand feet high.

Along this drive many hot spring and steam vents are seen. **Beryl Spring** is rather more than usually attractive and deserves particular notice, being the largest boiling spring in the canyon. It is fifteen feet across, and is located close by the roadside about a mile from the entrance to the canyon. The violent boiling of its surface, together with the hiss of escaping steam, causes some nervous apprehension to the feelings of the traveler, but strangely enough has no terror for the stage-horse, even though the road is almost constantly enveloped in steam.

Gibbon Falls, whose waters tumble in a foamy torrent down a series of steep cascades on one side of a bold rocky ledge, and on the other, stream in a thin shining ribbon of silvery spray from a height of over eighty feet, fittingly concludes the attractions of Gibbon Canyon.

From the falls the road descends to the Gibbon Canyon junction where are located the **Shaw & Powell** and **Wylie** noon stations. From this point the right hand road follows down the river to the

GIBBON FALLS—80 FEET.

Western Entrance, and the left hand road leads to the Lower Basin.*

For a distance of three or four miles from the junction, the route is over a succession of pine-clad terraces to the valley of the Firehole River, which unites with the Gibbon River to form the Madison, one of the principal sources of the Missouri River. This driveway is greatly admired, presenting as it does an ideal park-like appearance, all down timber and rubbish having been cleared away by the Road Department.

---

*See page 30 for continuation of trip from Northern Entrance.

ANIMALS OF YELLOWSTONE PARK.

A LTHOUGH unfenced, Yellowstone Park is perhaps the best game preserve in North America. Being suited to the habits of such a large number of species of large and small animals, it preserves them in their natural state free from the molestation of the hunter. With exception of the Mountain Lion and Coyote, both of which are very harmful to the

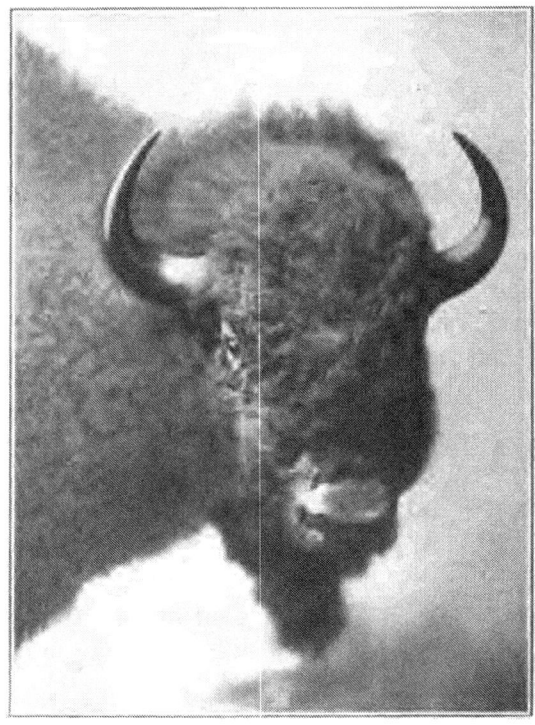

A BUFFALO HEAD.

THE TWIN CUBS.

young of the other large game, especially the young Mountain Sheep, Elk, Deer and Antelopes, all animals that naturally inhabit this remarkable region are protected in every possible way. All hunters and poachers are rigidly excluded, and in winter, when procuring forage is difficult, the Elk and Antelope are supplied with hay. On account of the fact that the buffalo is

A PARK BEAR.

fast becoming extinct throughout the country, a corral
has been constructed near Mammoth Hot Springs for
a small herd of these animals, where it is hoped they
will multiply and be perpetuated.

Of the bears that inhabit the park in great numbers,
the **Grizzly,** or **Silver Tip,** easily deserves first men-
tion; it is the most celebrated of all the bears in the
world. Although it is said that more hunters have
been maimed and killed by the Grizzly than all other
bears of the world combined, he seems to realize that
he is being protected and does very little harm in the
park. Unless he is cornered, or thinks he is cornered,
he will invariably flee from man. The high shoulders,
powerful proportions and grizzly-gray hair easily dis-
tinguish him from the others. He is a great traveler,
swims well, but is unable to climb trees; his food
consists of practically anything he can chew, but he is
decidedly partial to berries and fruits of all kinds.

BUFFALO NEAR FT. YELLOWSTONE.

The **Black Bear** is jet black all over except his nose, which is brown; however, a confusing fact about the Black Bear is that frequently its color runs into brown, or cinnamon colors. In one litter there have been found cubs both black and brown. When of a brown color it is called the **Cinnamon Bear**; both are smaller than the Grizzly, are good climbers and, though usually timid, fight in a rough and tumble fashion, with much roaring and growling.

The **Buffalo,** or **American Bison,** which but a few years ago grazed in countless thousands on the Western plains, are now counted in tens; only a score or more remain in their natural state—straggling remnants of perhaps the stateliest species of hoofed animal in America; these are roaming over secluded areas in the park unmolested and are seldom seen.

The **Prong-horned Antelope,** found only in North

ANTELOPE AND NORTHERN ENTRANCE ARCH.

America, lives in isolated bands in but few localities in the Rocky Mountains, chiefly in the Yellowstone Park. This keen-eyed animal, fleet of foot and timid, will doubtless soon become extinct in all places but the park; as it does not endure in captivity it must be preserved in its wild state. Like the Elk, Deer and

MOUNTAIN SHEEP.

Caribou, the Prong-horned Antelope sheds its horns each year, and they are renewed each year.

Big Horn Sheep, or **Mountain Sheep,** are found where the scenery is grandest in high mountain places where none but bold and reckless climbers would dare to go. Its young are reared in the highest and most inaccessible places, and as a result, the larger birds are their only dangerous enemy. Bands of

Mountain Sheep frequent the high bluffs overiooking Gardiner Canyon at the northern part of the park. They are also found in a few widely separated locali-

ELK IN HAYDEN VALLEY.

ties in the Rocky Mountains from British Columbia to Mexico. No other wild animal has circling horns; those of the Mountain sheep make nearly a complete circle and are built round and very heavy.

There are thousands of **American Elk**, or **Wapiti**, in Yellowstone Park, several photographs having been taken showing groups of several hundred. The Elk is as tall as a horse, handsomely formed, has a luxurious mane and imposing antlers. Even the young of this species are stately; they "step about with the air of a game cock." It seems remarkable that horns of such great size can be grown to maturity in a few months, to be lost and regrown each year. It is not uncommon for tourists to see Elk and Deer from the roadside while driving over the main highways of the park.

A PARK DEER.

A YOUNG ELK.

The **Deer** attracts fully as much if not more attention than the Elk on the part of the traveler; two members of the Deer family are prominent in the park, the **Black-tailed,** or **Mule Deer,** and the **White-tailed Deer.** The former has larger antlers, consisting of two Y's on each horn. The coat of the Black-tailed Deer is steel gray in winter and gray brown in summer. Except in the park it is being destroyed much faster than it breeds, which means an early extinction of this species. The White-tailed Deer, unlike the Mule Deer, is a skulker; it hides in the brush and carries its head low, so is seldom seen. Its name is derived from its long bushy tail, which is white underneath and pointed.

The most widely-known member of the cat family in

North America is the **Puma,** or **Mountain Lion;** it makes its den among the rocks or in the dense forests and preys upon every creature that can be killed and eaten, doing much harm to the young Elk, Deer, Mountain Sheep and Antelopes. The Mountain Lion is a good climber; it is tall for its weight, thin-sided and on an average about seven feet long from tip to tip. In color it is a brownish drab. On account of the diligent work on the part of the park authorities, this harmful animal is becoming practically extinct in the reserve.

Coyotes, like the Mountain Lion, prey upon the young of many valuable species; they, too, are "shot on sight" by the scouts and cavalrymen in the park. They are numerous in the lower altitudes of the park; not infrequently their dog-like yelping is heard in the vicinity of the hotels. Washouts and holes in the sides of ravines furnish dens for the coyote, which multiplies with comparative rapidity, having from five to seven puppies each year.

Of the small fur-bearing animals in the park, there are the Otter, Mink, Weasel, Marten, Skunk and Badger.

The **Otter,** being fond of water and living chiefly on fish, makes its home usually under the roots of a large tree overhanging the banks of a stream. It has webbed feet and a thick, flat tail for use in swimming. The fur of the Otter is very fine and of a dark brown color.

The **Mink,** unlike the Otter, is not aquatic, it preys on small mammals and fish when it can procure them, but lives chiefly on birds; it is smaller than the Otter, and its fur, which is yellowish or dark brown, is highly prized.

The **Common Weasel,** or **Ermine,** is a small, long-bodied animal with short legs, the smallest member of the Marten family. It kills grouse, ducks, rabbits and other animals, some ten times its own size, and is considered the most vicious of all animals. In summer its coat is brown, but white in winter, a striking manifestation of Nature's plan of protection.

The **Marten** lives on small rodents, birds and eggs, and spends so much time in the trees that it is often called the **Pine Marten.** Its habitat is on rugged and rocky forest-covered mountains, seldom in open country.

The **Common Skunk** is of conspicuous jet black color, with two wide stripes of white running lengthwise over its back; its fur is becoming valuable on account of the scarcity of Otter, Beaver, Mink and Marten; before being used, however, the white portions are dyed black.

The **Badger** has a broad, flat back, and like the Weasel, has very short legs and is very savage. It lives in burrows and feeds on squirrels and other ground game of every description. Along the park highways the **Tree Squirrel** is often seen, while the **Rock Squirrel** (Chipmunk) is likewise abundant. The **Ground Squirrel** lives in the open country in places like Swan Lake Valley, and is seldom seen in rocky places or in the trees.

The **Woodchuck,** or **Ground Hog** is much larger than any squirrel and is of a rich brown color. He is often seen by the roadside sunning himself near his burrow. In autumn he does not store up a winter's supply of provisions like the squirrel, but takes on a quantity of fat under the skin, then goes quietly to sleep in his burrow for four or five months when the winter is severest; hibernating like the bear.

The **Beaver** is celebrated for his engineering skill in building dams, some of great extent, for the purpose of providing in streams a safe refuge from its enemies. He constructs a water entrance to his house and a place below the freezing line for his winter supply of food. The Beaver is easily recognized by its broad, hairless tail, which it uses in swimming. It is not

A BEAVER HOUSE IN WINTER.

uncommon for Beavers to fell trees which are as much as a foot in diameter, by gnawing, and it is said that they cut them so they will fall toward their pond. The favorite bark prized by them in the park is the aspen. Beaver dams are seen from the roadway in Willow Park, in Beaver Lake at the foot of Obsidian Cliff, and in several other places in the reserve. The Beavers themselves are seldom seen during the daytime, or in fact at any other time; they work in the evening, beginning about an hour before sundown.

The **Muskrat,** chief member of the family of mice and rats in the park, is found along the banks of streams where burrows can conveniently be made. They are quite as at home in the water as Beavers, and like the Beavers, they have powerful tails which serve as the motive power in swimming. Muskrat fur when dyed a rich brown black, plucked and dressed, is known as "French seal."

Porcupines until recently have been abundant in the park; their disappearance leads to the theory that they have moved to other localities for some unexplained reason, rather than that they have suddenly become extinct.

They live chiefly upon bark and are equally at home in the tree-top or on the ground. It is known that the Porcupine has caused the death of more than one Mountain Lion and Lynx by means of its quills; any animal attempting to bite the Porcupine gets its mouth filled with spines, which prevent its eating, causing

death by starvation. It has been stated that the quills
are *thrown* by the Porcupine; this, however, is not the
fact. When attacked he huddles into a ball completely
covered with quills and strikes his adversary with his
tail, at the same time lodging in him many painful
spines.

Reptiles are rare in the park region, and it is a com-
forting fact that the Rattlesnake is not found above
6,000 feet altitude. The average altitude in the park
is 8,000 feet.

BIRDS OF YELLOWSTONE PARK.

OSPREY ALIGHTING.

While the variety of birds in Yellowstone Park is large, only a few of each kind are seen. The most important ones are the Eagle, Osprey, Sea Gull, Pelican, Vulture, Goose, Swan, Crane, Crow, Raven, Magpie, Lark, Bluejay, Blackbird, Robin, Grouse, Pheasant, and a large variety of ducks.

The Osprey, or Fish Hawk, usually builds its nest on inaccessible pinnacles and tree-tops near lakes and streams. The accompanying illustration shows an Osprey's nest in Gardiner Canyon; since the early days the rock pinnacle has had the misleading name of Eagle Nest Rock.

Wild ducks and geese are frequently seen from the roadways; and on Yellowstone Lake are many water fowl.

SEA GULLS AND PELICANS—YELLOWSTONE LAKE.

The view of the Sea Gulls and Pelicans was taken within sight of the Lake Hotel where there are usually a number of water-fowl.

"Large numbers of the Canada geese have reared their young in the park and showed little fear of molestation by visitors. Also ducks of many varieties. Pelicans and gulls occupy the entire surface of one small island in Yellowstone Lake as their nursery. More than seventy species of birds come to the park to rear their young."—General S. B. M. Young.

FISH AND FISHING.

The United States Fish Commission has had an important part in making Yellowstone Park one of the foremost resorts for the angler in America. With the exception of Yellowstone Lake and River, practically none of the streams or lakes had native trout, or fish of any kind, in their waters before the Commission stocked them. Since 1890 more than 100,000 fry have been planted in the various streams, and in 1904 a fish hatchery was built at the West Thumb of Yellowstone Lake.

In explanation of the lack of fish in this region, which seems so well suited to their habits, Mr. David S. Jordon in 1889 wrote as follows:

"The streams of the park are for the most part among the coldest and clearest of the Rocky Mountains, and apparently in every way suitable for the growth of trout . . . . yet, with exception of the Yellowstone itself, all these streams are destitute of fish life. The plateau is fringed with cataracts which no fish can ascend; each stream has a canyon and waterfall near the point where it exchanges the hard bed of lava for the softer rock below. So the best of trout streams for an area of 1,500 square miles are left without trout, because their natural inhabitants cannot get to them."

Today practically all of the streams in the reserve are well stocked, and afford excellent sport for the angler. Among the varieties of trout are: Rainbow, Brook, Loch Leven, Von Behr, and the native trout;

while in the Madison river, near the Western Entrance, are the Grayling, and in the Gardiner River
the White fish.

Regulations governing fishing prohibit the use of
any other means than the hook and line; no one person is allowed to catch more than twenty fish in one
day, and all fish under 8 inches in length must be returned to the water with the least damage possible
to the fish.

A BIG TREE NEAR MAMMOTH HOT SPRINGS AT AN ALTITUDE
OF OVER 6,000 FEET.

YELLOWSTONE TREES.

The forests which cover a large portion of Yellowstone Park are chiefly of one species, the **Black Pine** (*Pinus Murrayana*), sometimes called the Lodge pole pine on account of its proneness to grow high with very few branches. Over burnt areas it is the first to spring up; and it grows with comparative rapidity.

Next in importance is the **Balsam** *(Abies subalpina),* found to large extent on steep slopes and in moist places, flourishing near the snow fields. It is considered the most beautiful tree in the park forests.

The **White Pine** *(Pinus flexilis),* unlike the balsam, flourishes best in the lower altitudes. It is a hardy but not especially ornamental tree; specimens are seen along the Gardiner River and in the vicinity of Mammoth Hot Springs.

The **Cedar** (*Juniperus scopulorum*), is seen near Mammoth Hot Springs. It is **extremely slow-growing, and while of little commercial value, it is attractive on account of its ancient, gnarled appearance.**

Another species of cedar which is common throughout the park is in appearance more like a shrub than a tree; the *Juniperus sibirica.* It is a rich green in color, grows close to the ground and spreads in all directions from the center.

Other trees of less importance are the Dwarf Maple, Quaking Aspen, Willow and Alder.

Forest growths in the park are for the most part stunted; and are of little value as lumber, although the black pine is used extensively for poles and fuel, the latter use being made of the dead and down timber, which is abundant.

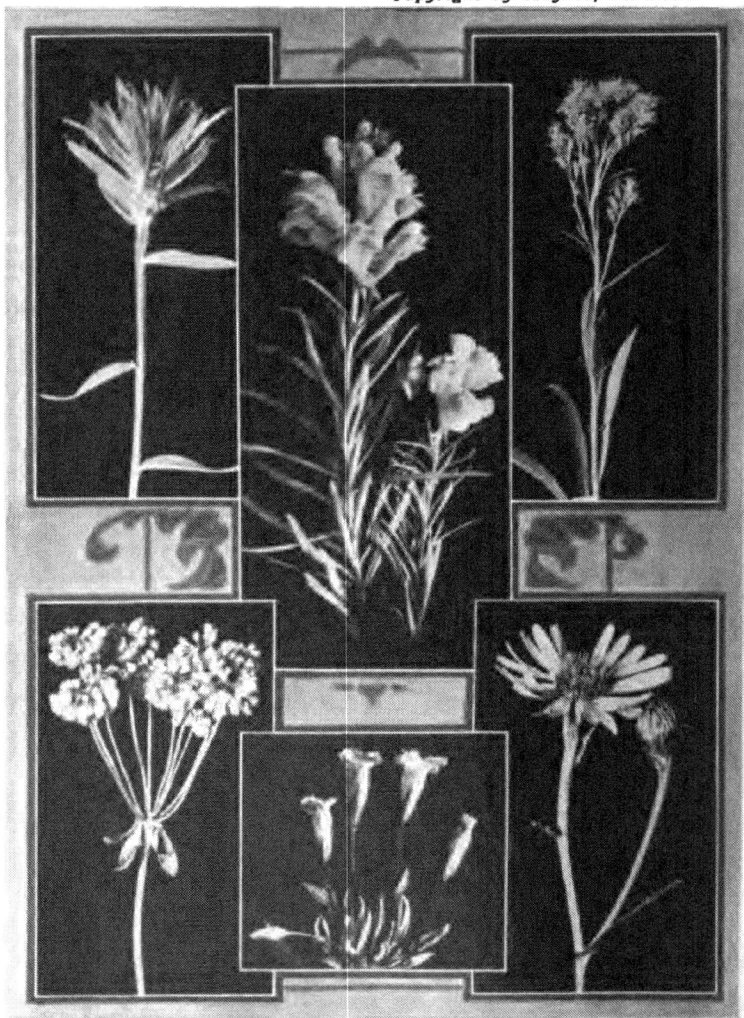

PAINT BRUSH,	BUTTER AND EGGS,	GOLDEN ROD,
Castilleia miniata.	*Linaria vulgaris.*	*Solidago Corymbosa.*
WILD BUCKWHEAT,		ASTER,
Eriogonum	GENTIAN,	*Machaeranthera*
umbellatum.	*Gentiana elegans.*	*Bigelovii.*

FLOWERS OF YELLOWSTONE PARK.

MONKEY FLOWER,
Mimulus Langsdorfii.

Yellowstone flowers, occurring as they do in almost countless varieties, and in forms frequently quite different from those customary in lower altitudes, afford exceptionally good material for botanical study.

"A plant is not to be studied as an absolutely dead thing, but rather as a sentient being. . . . . Since man has learned that the universal brotherhood of life includes himself as the highest link in the chain of organic creation, his interest in all things that live and move and have a being has greatly increased. . . . He sees in each of the millions of living forms with which the earth is teeming, the action of many of the laws which are operating in himself; and has learned that to a great extent his welfare is dependent on these seemingly insignificant relations; that in ways undreamed of a century ago they affect human progress."—Clarence Moores Weed.

One of the most beautiful flowers of the region is the **Fringed Gentian** *(Gentiana elegans)*, which grows in profusion in the low, moist meadows and in the vicinity of the geysers. Although usually of a beautiful blue color, specimens have been found in the park which are pure white; these being highly prized by collectors. The Gentian has been chosen for the state flower of Wyoming; its name is from *Gentius,* King of Illyria, who is credited with having first discovered its medicinal virtue.

The state flower of Montana is the **Bitter-Root** *(Lewisia rediviva),* which gives the name to the Bitter-root Mountains and river. It grows abundantly on the hills in the vicinity of Mammoth Hot Springs and flowers in June and July. The flower grows close to the ground and is of a delicate pink color. Its roots, which are fleshy and farinaceous, have been used extensively for food by the Indians. The name *Lewisia* is in honor of Capt. Lewis of the famous Lewis and Clark Expedition.

The **Evening Primrose** *(Oenothera muricata)* is usually found in dry localities, as in Golden Gate Canyon and Snow Pass; although white, or pale yellow, at first, it later turns a delicate rose color and is very fragrant. It has four delicate, spreading petals and is about two and one-half inches across; the blossoms appear only in the evening and lie close to the ground.

The true **Forget-Me-Not** *(Mysotis alpestris)* grows only in the higher altitudes in the park, although similar flowers are common throughout the region; along the Yellowstone River and on the sides of Mt. Washburn it is very common, growing in thick clusters close to the ground. Its color is pale blue usually, though in some places it is very dark. The name is from the words *mouse* and *ear*, due to the fact that in some species the leaves are short and soft.

The **Harebell** *(Campanula rotundifolia)* grows in the moist, rocky places along the roads, and in the uplands, being quite common in the park. Its bell-shaped flowers of a delicate blue adorn the tips of very slender stems; it blooms from June until September. The name *Campanula* is a diminutive of the Italian *campana,* a bell.

The **Shooting Star** *(Dodecatheon meadia)* grows on moist, rocky places along the roads, in the open woods, and prairies of the park. In color it is a purplish-pink, sometimes white, and seems appropriately named, as the flowers nod with petals bent backward as if the flower were really darting through the air.

The **Larkspur** *(Delphinium,* several species*),* is quite abundant; it grows in open deciduous woods and prairies, is of dark blue color, and is popular in bouquets. This plant is considered poisonous to cattle and horses; its name *Delphinium* is from *Delphin* in allusion to the shape of the flower, which is not unlike the classic dolphin.

The *Mentzelia decapetala,* a rare, night-blooming flower of exquisite beauty, grows in the vicinity of Mammoth Hot Springs. The average specimen is four inches across, with ten petals, of a pale yellow color. Another species having five petals is found here, but less commonly. A peculiarity of these plants is their long barbed leaves, which cause the flower to stick to one's coat without other means. Locally the *Mentzelia* has been erroneously called Night-Blooming Cereus.

The **False Dragon Head** *(Physostegia virginiana)*, has large, rose or flesh-colored blossoms, which are showy, in general appearance resembling the False Fox-Glove. Its foliage is of a dark, glossy, green color, and it grows in the moist places near the streams and geysers.

The **Ground Phlox** *(Phlox subulata)*, grows in many places along the roads, its habitat being in dry, rocky and sandy places. In color the Phlox is found both pink and white; several species occur in the park. The flowers are small, but grow in clusters over a bed of green close to the ground, producing a very striking effect.

The **Lupine** *(Lupinus perennis)*, is very common. It is usually a deep purplish-blue, rarely white. Its habitat is in dry, sandy soil, where it grows abundant-ly. Lupine is derived from *lupus,* a wolf, because these plants were thought to devour the fertility of the soil, while as a matter of fact they seem to prefer the less fertile spots.

The **Columbine** *(Aquilegia canadensis)*, is consid-ered one of the most exquisite flowers. It has been selected state flower of Colorado. The flowers are red outside and yellow within, and are large and showy.

They are found in many sections of the park, in localities which are forested and rather high in altitude, as in the neighborhood of Mt. Washburn, Undine Falls and Bunsen Peak.

The **Painted Cup** *(Castilleja coccinea),* is usually an intense red, rarely yellow, and looks as though it had been just dipped in paint. It flourishes in shady, sandy places frequently in grassy patches, where its brilliant color is in marked contrast to the green.

A curious flower which may be dried and still preserves its apparent freshness indefinitely is, the **Everlasting** *(Antennaria diocia rosea)* of a pink, occasionally white color. It occurs in the vicinity of Mammoth Hot Springs and Yellowstone Lake.

Buttercup *(Ranunculus montansis),* a pretty yellow flower, blooms in June and July and is found near the Grand Canyon and Yellowstone Lake.

Umbrella Plant *(Eriogonum subalpinum)* occurs in several species throughout the park and blooms the greater part of the summer.

Dogtooth Violet *(Erythronium grandiflorum)* grows in the rich wet soil in the neighborhood of Swan Lake, in the open woods and thickets, and near the streams. The flower has six yellow long, pointed petals and is about two inches across. The stem is not leafed.

Other flowers of less importance in the park are the yellow pond lily, golden rod, clematis, ox-eye daisy, dandelion and late purple aster.

HISTORICAL.

ALTHOUGH part of it was included in the great Louisiana Purchase of 1803, the Yellowstone Park was not then known to white men. Probably the first one who ever saw any of its hot springs or geysers was John Colter who left the celebrated Lewis and Clark Expedition, which was on its return to St. Louis, in 1806 and started for the head waters of the Missouri River to trap and hunt. This lone adventurer passed northward in 1807 from the mouth of the Big Horn to the Forks of the Shoshone River where he discovered an immense tar spring; he continued on through a country where much hot spring and geyser phenomena exist and down the Yellowstone River to the ford at Tower Falls, thence out near the northeastern corner of what is now the National Park.

After four years of peril among the Indians and a miraculous escape from the hostile Blackfeet, he returned in 1810 to St. Louis. His wonderful tales were hard to believe and the place he described, (which was thought to be the product of his imagination), was termed "Colter's Hell."

The Park had been described in part by some of the early hunters, but their knowledge of the place was limited, due to the fact, no doubt, that the region was so difficult to explore; and it is a fact worthy of note that until 1834, no written description of these geyser regions had ever appeared. But in that year, one W. A. Ferris visited the Upper and Lower Geyser Basins and prepared a description of what was there. The next written account of the region appeared ten years later based on information furnished by the noted Rocky Mountain Guide, James Bridger, "He (Bridger) gives a picture most romantic and enticing of the head waters of the Yellowstone," to quote from Gunnisen's History of the Mormons, "A lake, sixty miles long, cold and pellucid, lies embosomed among high precipitous mountains. On the west side is a sloping plain, several miles wide, with clumps of trees and groves of pine. The ground resounds with the tread of horses. Geysers spout up seventy feet high, with terrific hissing noise, at regular intervals. Waterfalls are sparkling, leaping and thundering down the precipices, and collect in the pool below. The river issues from this lake, and for fifteen miles roars through the perpendicular canyon at the outlet. In this section are the 'Great Springs,' so hot that meat is readily cooked in them, and as they descend on successive terraces, afford at length delightful baths. On the other side is an acid spring, which gushes out in a river torrent; and below is a cave, that supplies 'vermillion' for savages in abundance." Probably no other man in Bridger's time had such a comprehensive knowledge of the Park region.

Captain John Mullan mentions the Park geysers in his report to the government in 1853 and states that he visited them.

Colonel Raynold's Expedition could not penetrate the region when it attempted to explore it in 1860, on account of the snow encountered; the party encircled it however and learned much from the tales of hunters and trappers who had visited it. Colonel Raynold in his report on the "Exploration of the Yellowstone" in

1859-60 states regarding the "Munchausen Tales" about the Park:

"One was to this effect: 'In many parts of the country petrifactions and fossils are very numerous, and, as a consequence, it was claimed that in some locality (I was not able to fix it definitely) a large tract of sage is perfectly petrified, with all the leaves and branches in perfect condition, the general appearance of the plain being like that of the rest of the country, but all is stone; while the rabbits, sage hens and other animals usually found in such localities are still there, perfectly petrified, and as natural as when they were living; and, more wonderful still, the petrified bushes bear the most wonderful fruit; diamonds, rubies, sapphires, emeralds, etc., etc., as large as black walnuts, are found in abundance.' "

The following is taken from the report made to the late Dr. F. V. Hayden, chief of Geological Survey of Territories, by Henry Gannet, E. M., on the geographical field work of the U. S. Geological Survey during the season of 1878:

"The story of the remarkable fruit born by these stone trees is not far from correct, the main difference between the story and the fact being that the fruit is borne on the outside and inside of the trunk of the trees, instead of on the ends of the branches. The mineral species are not as given in the story, either, but this is a matter of no vital importance. In the process of the silicification of wood the last result of all is the production of quartz crystals. The trunk is converted totally into crystalline quartz, radiating from within outward, the crystals being all crowded out of shape. The inside and outside of the hollow cylinder of quartz, which represents the former tree, are covered with the characteristic quartz pyramids. Such products of silicification are very abundant in the Park, particularly on Amethyst Ridge, and are, undoubtedly, the stone fruit of the petrified trees and bushes. The crystals are colorless, amethystine or yellow, and according to the color, are known to the mountain man as diamonds, amethyst, topaz, etc. It is unnecessary to say that the part of the story relating to animal life was manufactured from the whole cloth.

"In 1863, Captain W. W. DeLacy, in command of a large party of prospectors, left Montana to prospect on the upper waters of the Snake. Striking that river near the junction of Henry's Fork, they followed up the main river through the canyon, prospected in Jackson's Hole, and, not finding gold in paying quantities, they broke up the party, some returning one way, some another. Captain DeLacy, with a portion of the party, followed up the Snake and Lewis Fork, discovering Lewis and Shoshone (DeLacy's) Lakes, the Shoshone and the Lower Basins. The geographical work done by Captain DeLacy on this trip was embodied in a map of Montana, drawn by himself, and published by authority of the territory in 1864-65, and the material thus made public was afterwards used by the land office in the compilation of maps of that region.

"The results of this trip seem to have attracted little or no attention, for we hear of no one going into the country until 1869, when the two prospectors, Cook and

Folsom, made a prospecting tour through the Park. They followed the Yellowstone up to the mouth of the East Fork, then up the latter stream for a few miles, crossing over to the Yellowstone at the Great Falls; thence they went up this stream to the foot of the lake and around the east side of the latter to the extremity of the west arm; thence crossing over to Shoshone Lake and Lower Geyser Basin on the Madison or Firehole, and finally left the country by following down the Madison River.

"Their story, written by Mr. David E. Folsom, and published in the Chicago Western Monthly for July, 1870, immediately attracted attention, and the following summer a large party, composed of prominent citizens of Montana, under the leadership of General Washburn, then Surveyor General of Montana, was made up for the purpose of exploring this region. Among the party was Hon. N. P. Langford, first superintendent of the Park; Hon. Cornelius Hedges, who first proposed setting apart this region as a National Park; Hon. T. C. Everts and S. T. Hauser, accompanied by a small escort from Fort Ellis in charge of Lieut. G. C. Doane.

"This party made quite extensive explorations on the Yellowstone and Madison rivers. Passing up the Yellowstone by the well-known trail, they traveled completely around the lake, visiting all localities of interest along the route, with the single exception of Mammoth Hot Springs, on Gardiner River."

Many of the most prominent features of the Park were named by this party—Mount Washburn the famous promontory, Old Faithful Geyser, the Castle,

Beehive, National Park Mountain, and several other conspicuous points of interest.

While they were near Yellowstone Lake, Mr. Everts strayed away from the party and was lost in this almost impenetrable country. After making a diligent but unsuccessful search for him, the party was forced to continue their journey; and when they returned, finding that Mr. Everts had not been heard from, two men with provisions and ammunition were immediately sent out in search of him.

In the meantime Mr. Everts had been overtaken by a severe storm and while searching on foot for evidence of a trail, lost his eye glasses and was unable to return to his horses. Three weeks later he was found by Jack Barronette in a starved and half demented condition crawling on his hands and knees. Happily he fully recovered from his unfortunate experience.

The success of the Washburn Expedition and the accounts furnished by its members led to extensive explorations in 1871.

Expeditions under Dr. F. V. Hayden of the United States Geological Survey, and Captains Barlow and Heap of the Engineer Corps of the Army resulted in the discovery of Mammoth Hot Springs and the route from the Lower Basin to the Yellowstone River. A map of the outline of the Yellowstone Lake was made, and collections of specimens were gathered throughout the region. The reports which followed were scientific and very complete.

Until 1872, the region was open to settlers and there

were no restrictions on hunting, trapping, gathering
specimens and the fencing-in of the geysers for private
gain. But these dangers were foreseen and the region
was set aside as a National Park, March 1, 1872, when
President U. S. Grant affixed his signature to the Act
of Dedication.

THE ACT OF DEDICATION OF YELLOW-STONE NATIONAL PARK.

Approved March 1, 1872.

*Be it enacted by the Senate and House of Representatives
of the United States of America in Congress Assembled:*

That the tract of land in the Territories of Montana
and Wyoming, lying near the headwaters of the Yellowstone River, and described as follows, to wit:
Commencing at the junction of Gardiner River with
the Yellowstone River, and running east to the meridian passing ten miles to the eastward of the most
eastern point of Yellowstone Lake; thence south
along the said meridian to the parallel of latitude passing ten miles south of the most southern point of
Yellowstone Lake; thence west along said parallel
to the meridian passing fifteen miles west of the most
western point of Madison Lake; thence north along
said meridian to the latitude of the junction of the
Yellowstone and Gardiner Rivers; thence east to place
of beginning—is hereby reserved and withdrawn from
settlement, occupancy or sale under the laws of the
United States, and dedicated and set apart as a public
park or pleasure ground for the benefit and enjoyment
of the people; and all persons who shall locate, settle

upon or occupy the same or any part thereof, except as hereinafter provided, shall be considered trespassers and removed therefrom.

Sec. 2. The said public Park shall be under the exclusive control of the Secretary of the Interior, whose duty it shall be, as soon as practicable, to make and publish such rules and regulations as he may deem necessary and proper for the care and management of the same. Such regulations shall provide for the preservation from injury or spoliation of all timber, mineral deposits, natural curiosities or wonders within said park and their retention in their natural condition.

The Secretary may, in his discretion, grant leases for building purposes, for terms not exceeding ten years, of small parcels of ground, at such places in said park as shall require the erection of buildings for the accommodation of visitors; all the proceeds of said leases, and all other revenues that may be derived from any source connected with said park, to be expended under his direction, in the management of the same, and the construction of roads and bridle paths therein. He shall provide against the wanton destruction of the fish and game found within said park, and against their capture or destruction for the purpose of merchandise or profit. He shall also cause all persons trespassing upon the same after the passage of this act to be removed therefrom, and generally shall be authorized to take all such measures as shall be necessary or proper to fully carry out the objects and purpose of this act.

PROF. ARNOLD HAGUE'S CAMP—U. S. GEOLOGICAL SURVEY.

In 1873, Captain W. A. Jones, U. S. A., took a large party through the Park. He entered it from the head of the Stinking Water, crossing one of the many passes near Mt. Chittenden. After visiting most of the points of interest in the Park, he went out via the Upper Yellowstone, on the way verifying the old trapper's legend about the 'Two Ocean River,' and discovering a practical pass (Togwotee Pass) and route from the south to the Park. This discovery was by far the most valuable result of the expedition.

In 1875 Captain William Ludlow, U. S. A., in charge of a reconnaissance in Central Montana, made a flying trip to the Park. He developed little that was new save accurate measurements of the Upper and Lower Falls of the Yellowstone.

. General O. O. Howard crossed the Park in his famous pursuit of the Nez Percé Indians in 1877; the year that Col. P. W. Norris was made Superintendent to succeed Hon. N. P. Langford who had held that office five years. Mr. Langford did more for the Park than can be reckoned; he served as Superintendent without pay or remuneration of any kind and had upheld the "National Park Idea" from the time the Expedition of 1870 talked of the plan until the Act of Dedication was finally passed in 1872.

The United States Geological Survey resumed work in the Park in 1878 under Dr. F. V. Hayden; and in 1883 a report was published giving detailed descriptions of the points of interest, as well as scientific discussions of the phenomena observed. This report is beautifully illustrated with color-plates, engravings, diagrams and maps.

In August, 1883, President Arthur with the Secretary of War, Lieutenant-General Sheridan of the Army, Senator Vest and several other distinguished officers and civilians visited the Park in the most elaborate pack-train expedition that has ever been enrolled. The route lay from Green River on the Union Pacific R. R., to Livingston on the Northern Pacific Railway.

Mr. F. Jay Haynes, Official Photographer of the Park, who accompanied the party, procured many interesting photographs on this famous expedition.

In 1886 the administrative policy of the Park was changed. Capt. Moses E. Harris of the military was appointed Superintendent of the Park to succeed Col. D. W. Wear, civilian; and since that time all the superintendents have been officers in the Army.

PRESIDENT ARTHUR'S PARTY AT UPPER BASIN, AUGUST, 1883.

Standing—Reading from left—Col. Mike Sheridan, U. S. A., Gen. Anson
   Stager, Capt. Philo Clark, U. S. A., Judge Rawlins, Col. J. F. Greg-
   ory, U. S. A.

Sitting—Reading from left—Gov. Schuyler Crosby, Mont., Gen. P. H. Sheridan, U. S. A., President C. A. Arthur, Secy. of War Robt. T. Lincoln,
   Senator Geo. G. Vest.

Winter Exploration in 1887.—In January, 1887, the
first successful exploration of the Yellowstone region
was made. Lieutenant Fredrick Schwatka of Arctic
fame headed the party consisting of F. Jay Haynes,
photographer, several eastern gentlemen and a corps
of guides, packers and assistants. Their outfit consisted of astronomical instruments, photographic equipment, sleeping bags and provisions which were drawn
on toboggans; the party used Norwegian skis and
Canadian web snowshoes, but the snow was so light
that they sank readily and the toboggans were exceedingly difficult to draw. It took three days to cover

the twenty miles from Mammoth Springs to Norris Basin; and the temperature the first night at Indian Creek was 37° below zero.

Unfortunately Lieut. Schwatka fell ill at Norris and was unable to proceed.   Mr. Haynes, desirous of obtaining a collection of winter scenes of Yellowstone Park, employed two of the sturdiest men of the Schwatka Party, and with Edward Wilson, a government scout, resumed the journey.

The toboggans were abandoned and this party resorted to the customary fashion of packing their equipment and provisions on their backs—each man carrying about forty-five pounds.

Norris Basin was a grand sight.   Craters heretofore

PHOTOGRAPHER HAYNES, PARK MID-WINTER.

unnoticed by these men familiar with the Park in summer, steamed copiously. The foliage was heavily laden with ice near the steam vents and geysers, producing all the fantastic forms possible to imagine; while the entire basin resembled a vast manufacturing centre.

CAMPED FOR THE NIGHT.

Tall trees buried in the snow appeared like bushes, sage brush and bowlders were entirely hidden, and the general aspect of the country was completely changed; the average depth of the snow being about eight feet.

The steam rising fully two thousand feet from the geysers in the Upper Basin could be seen from the Lower Basin, nine miles away. The Upper Basin presented the most striking appearance; the greater quantity of steam and more numerous active geysers presented an increased variety of peculiar effects.

The beautifully colored walls of the Grand Canyon were masses of pure white. The north half of the Great Falls hung in immense icicles 200 feet in length. An ice bridge fully 100 feet high was formed at the

SNOW-SHOE PARTY IN HAYDEN VALLEY.

base of the falls coming up fully to the spray line (about one-third the height of the falls). The brink was frozen over and was hidden in an arch of ice fully a dozen feet thick.

Thousands of elk were seen on the exposed ridges of Mt. Washburn, this being their winter range. The trip over Mt. Washburn was one of hardship and unusual privation; a blinding snowstorm overtook the little party of four which lasted three days; during which time they wandered day and night without shelter, provisions or fire. After the storm abated, they found their way to Yancey's exhausted after an adventure which nearly cost them their lives.

The circuit covered was about 200 miles, and the thermometer ranged from 10° to 50° below zero during the twenty-nine days consumed in making the trip.

·The Park is an ideal resort in summer, but it is not likely to become a winter resort on account of the extreme rigors of its climate.

Winter Expedition to the Game Ranges in 1894.— Early in March, 1894, a party was organized at Fort Yellowstone for the purpose of visiting the winter ranges of the game, to ascertain, as near as possible, the exact number of buffalo that still exists, and secure photographs of the same. The party consisted of Captain Scott, Lieut. Forsyth, Scout Burgess, Mr. Burns, Photographer Haynes and three non-commissioned officers. Mounted on the Norwegian snowshoes, with packs of sleeping bags, provisions and camera, they proceeded directly to Hayden Valley *via* Norris and the Grand Canyon. As most of the Buffalo congregate there during the winter months, they found eighty-one buffalo in the valley, seventy-three comprising the main herd, and numerous small groups of elk aggregating fully 300. After a stay of several days in

Hayden Valley the party went to Yellowstone Lake. Captain Anderson, Superintendent of the Park, had instructed Scout Burgess not to overlook the country east of the lake, as a small herd of buffalo usually winter there. The first day out from the lake only elk were seen by the scout and his companion, there being no sign of buffalo. They went into camp about twelve miles up Pelican Creek.

The second day they discovered, in a secluded spot, the "cache" of a poacher, very much to their surprise, as it was supposed that no one was in the Park killing game. The "cache" consisted of a canvas tepee, sleeping bag, provisions and toboggan and six buffalo heads suspended in a tree near by. A trace of fire in the tepee led the scout to believe that the poacher was in the vicinity, and to capture him, was the next move. It had been snowing constantly all the morning, and all snowshoe tracks leading from the camp were entirely obliterated. Some five miles from the camp they heard five or six rifle shots fired in rapid succession. Hastening through the timber to the opening in the direction of the firing, they came directly upon the poacher. He had driven six of the buffalo into the deep snow and slaughtered the entire band. Knowing these men to be of a desperate character, and being armed only with a pistol, it was a brave act for Scout Burgess to arrest him. Fortunately it was snowing hard, and the approach of the scout was not noticed by the poacher or his faithful dog until the arrest was made. He was taken to the Lake Hotel and escorted from there to the guard house at Fort Yellowstone.

Besides the twelve buffalo that were killed by this poacher, a small herd of seven were seen in the Pelican country, making less than 100 now in existence. If these can be protected they will increase rapidly, otherwise the only remaining species of large American game (the bison) will soon be exterminated. Elk were seen in the foothills of Mt. Washburn, on Specimen Ridge, along the east fork of the Yellowstone, on Slough Creek and along the Yellowstone to Mt. Everts, in great numbers. Fully 5,000 wintered in the above localities. Small bands of mountain sheep, deer and antelope were seen on Mt. Everts. The open water of the Yellowstone between the lake and falls was alive with duck and swan. The red fox and coyote were numerous, and an occasional black fox and footprints of mountain lion and bear were seen. The party was in the Park about thirty days and traveled over 300 miles.

Visitors to the Park each summer are numbered in thousands; they come from all over the world to see wonders that a few years ago no one but the mountain man, had ever heard of. The Park is growing in popularity each year, and justly so, for where are geysers that can compare to Old Faithful, the Giant, and a thousand others? Is there a pool elsewhere to compare with the Morning Glory? A falls like the Great Falls or a canyon that has both grandeur and beautiful coloring combined, like the Grand Canyon?

SECRETARIES OF THE INTERIOR

Since the Act of Dedication, March 1, 1872.

NAME	From	Date of Commission	Administration
Hon. Columbus Delano	Ohio	Nov. 1, 1870	Pres. Grant
Hon. Zachariah Chandler	Mich	Oct. 19, 1875	Pres. Grant.
Hon. Carl Schurz	Mo	Mar. 12, 1877	Pres. Hayes.
Hon. Samuel J. Kirkwood	Iowa	Mar. 5, 1881	Pres. Garfield and Arthur
Hon. Henry M. Teller	Colo	Apr. 6, 1882	Pres. Arthur.
Hon. Lucius Q. C. Lamar	Miss	Mar. 6, 1885	Pres. Cleveland.
Hon. William F. Vilas	Wis	Jan. 16, 1888	Pres. Cleveland.
Hon. John W. Noble	Mo	Mar. 6, 1889	Pres. Harrison.
Hon. Hoke Smith	Ga	Mar. 6, 1893	Pres. Cleveland.
Hon. David R. Francis	Mo	Sept. 1, 1896	Pres. Cleveland
Hon. Cornelius N. Bliss	N. Y.	Mar. 5, 1897	Pres. McKinley
Hon. Ethan A. Hitchcock	Mo	Dec. 21, 1898	Pres. McKinley & Roosevelt.
Hon. James R. Garfield	Ohio	Jan. 15, 1907	Pres. Roosevelt.
Hon. Richard A. Ballinger	Wash	Mar. 5, 1909	Pres. Taft.
Hon. Walter L. Fisher	Illinois	Mar. 13, 1911	Pres. Taft
Hon. Franklin K. Lane	Calif.	Mar. 5, 1913	Pres. Wilson

SUPERINTENDENTS OF YELLOWSTONE
PARK FROM 1872 TO 1916.

APPOINTED FROM CIVIL LIFE.

N. P. Langford............................May 10, 1872 to April 18, 1877
Philetus W. Norris........................April 18, 1877 to Feb. 2, 1882
Patrick H. Conger.........................Feb. 2, 1882 to July 28, 1884
Robert E. Carpenter.......................Aug. 4, 1884 to May 29, 1885
David W. Wear.............................May 29, 1885 to Aug. 1, 1886

ARMY OFFICERS DETAILED FOR DUTY AS ACTING SUPERINTENDENTS

Capt. Moses Harris.......5th Cav., U. S. A...Aug. 17, 1886 to May 31, 1889
Capt. F. A. Boutelle......1st Cav., U. S. A...June 1, 1889 to Feb. 14, 1891
Capt. Geo. S. Anderson....6th Cav., U. S. A...Feb. 15, 1891 to June 22, 1897
Col. S. B. M. Young......3rd Cav., U. S. A...June 23, 1897 to Nov. 15, 1897
Capt. James B. Erwin.....4th Cav., U. S. A...Nov. 16, 1897 to Mar. , 1899
Capt. W. E. Wilder.......4th Cav., U. S. A...Mar. , 1899 to June 22, 1899
Capt. Oscar J. Brown.....1st Cav., U. S. A...June 23, 1899 to July 23, 1900
Capt. Geo. W. Goode.....1st Cav., U. S. A...July 24, 1900 to May 7, 1901
Capt. John Pitcher.......1st Cav., U. S. A...May 8, 1901 to May 13, 1907
Gen. S. B. M. Young......U. S. A., Retired....May 14, 1907 to Nov. 27, 1908
Maj. H. C. Benson.......14th Cav., U. S. A...Nov. 28, 1908 to Sept. 29,1910
Col. L. M. Brett,........1st Cav., U. S. A....Sept. 30, 1910 to

OTHER PERSONS PROMINENT IN PARK AFFAIRS.

Anderson, Ole A., commercialized the practice of coating various articles in the hot water at Mammoth Springs, and conducted this unique business many years.

Arthur, Pres. Chester A., was the first president to visit the Park. He made the memorable trip in 1883 with General Sheridan and party from the Union Pacific R. R. through the Park to the Northern Pacific R. R.

Barlow, Capt. J. W., of the Engineers Corps U. S. Army led one division of the Expedition of 1871 when the Mammoth Hot Springs was discovered.

Baronett, C. J., scout and guide, built the first bridge across the Yellowstone in 1871 east of Yancey's. Mr. Everts lost from the Washburn-Langford Expedition in 1870 was found by him; Baronett Peak commemorates his name.

Bassett Bros., operated the first tourist line in the early 80's through the Park at the Western Entrance, from Beaver Canyon, Idaho, on the Utah Northern R. R.

Bridger, James, who lived from 1804-81, was the leading spirit in the Rocky Mountain Fur Company, the discoverer of Great Salt Lake, and the best informed man of his time on the Yellowstone Park.

Brothers, Henry J., completed the "Old Faithful Geyser Baths" at Upper Basin in 1915, which has a pool 100x50 feet in size.

Bryant, R. C., conducted camping parties through the Park from the Western Entrance for several years, and in 1912 sold out to Shaw & Powell.

Bunsen, Robt. W., after whom Bunsen Peak was named, invented the Bunsen Burner, Alkalimeter and many other scientific instruments; and advanced the most authoritative explanation of geyser action.

Burgess, Scout J. C., in the winter of 1894 captured a notorious poacher in the Park. This capture led to the passage of the present strict laws protecting the wild animals in Yellowstone Park.

Child, Harry W., now president of the Yellowstone Park Transportation Company and the Yellowstone Park Hotel Company has been in the Park many years. The present high standard of hotel service and transportation is due in la large measure to him.

Child, Huntley, is general manager of the Yellowstone Park Hotel Company.

Chittenden, Col. H. M., of the Corps of Engineers U. S. Army had charge of all the roads and bridges in the Park, for a number of years prior to 1904. He constructed the Stone Arch at the Northern Entrance, the concrete viaduct in Golden Gate and the road through the Hoodoos; he built the steel and concrete arch bridge near the Canyon, and the road to the summit of Mt. Washburn—engineering achievements of the highest order.

Clagett, Hon. Wm. H., delegate from Montana, introduced the Park Bill in the House, Dec. 18, 1871, and Senator Pomeroy of Kansas in the Senate.

Colter, John, a private in the famous Lewis and Clark Expedition was the first white man to visit Yellowstone Park. In 1807 he entered the Park at the southwest and left the region near the northeast corner.

Comstock, Prof. T. B., geologist of Captain Jones Expedition in 1873; the Comstock theory of geyser action appeared in his report of that year.

De Lacy, Capt. W. W., headed a prospecting expedition from Virginia City, Aug. 3, 1863, and discovered the pass between Shoshone Lake and the Upper Basin.
De Lacy Creek was named in his honor.

De Smet, Father, Jesuit missionary, in 1852 gave the first correct location of Yellowstone Park: Between the 43d and 45th degrees of latitude and the 109th and 111th degrees of longitude.

Doane, Lieut. G. C., was in command of the escort of the Washburn-Langford Expedition of 1870. His descriptions of the Canyon and other portions of the Park are most vivid and interesting. In 1875 he escorted Secretary of War Belknap and other distinguished gentlemen through the Park.

Dunraven, Lord, visited the Park in 1874. Dunraven Peak near Mt. Washburn was named after this enthusiastic writer.

Everts, Truman C., member of the Washburn-Langford Expedition, was lost from the party in crossing the continental divide between Yellowstone and Heart Lakes. He wandered 37 days until found by Scout Baronett a few miles from Mammoth Hot Springs. Mt. Everts commemorates his name.

Ferris, W. A., of the American Fur Company, visited the Upper and Lower Basins with two Indians in 1834 and wrote the first authentic descriptions of the geyser basins.

Folsom, W. A., with two companions, C. W. Cook and Wm. Peterson, made a thirty-six day expedition into the Park from Montana in 1869. His account of the trip appeared in the *Western Monthly*, Chicago, in July, 1871.

Gibbon, Gen. John A., commanded an expedition into the Park in 1872. Gibbon Canyon and Falls were named in his honor.

Gibson, Hon. Sir Charles, was first president of the Yellowstone Park Association. Virginia Cascades was named for his daughter.

Grant, Pres. U. S., signed the Act of Dedication, Mar. 1, 1872, creating the Yellowstone National Park.

Hague, Prof. Arnold, of the U. S. Geological Survey compiled a scientific report on Yellowstone Park published in 1899.

Hamilton, C. A., in 1915 succeeded H. E. Klamer as proprietor of the curio store at Upper Basin.

Hatch, Rufus, of New York, secured the first franchise for hotel improvements in the Park-1883. He constructed the Mammoth Hotel which was opened in 1884 and later sold to the Yellowstone Park Association.

Hayden, Dr. F. V., has been identified wth the Park as geologist since 1859. The Hayden U. S. Geological Survey Expedition of 1871 resulted in great benefit to the Park. The valley between Yellowstone Lake and the Canyon bears his name.

Haynes, F. Jay, president of the Yellowstone-Western Stage Company since its organization in 1898 and the Cody-Sylvan Pass Motor Co. He came to the Park in 1881 as photographer, accompanied President Arthur and General Sheridan on their trip in 1883, and made two winter trips of the Park, 1887 and 1894.

Haynes, J. E., in 1916 became official photographer of the Park, succeeding F. Jay Haynes, who was the official photographer for over thirty years.

Hedges, Cornelius, member of the Washburn-Langford Expedition. While in camp at the junction of the Gibbon and Firehole rivers, Sept. 19, 1870, first suggested to the party that the region be set aside as a National Park.

Henderson, G. L., brother of Speaker Henderson of Iowa, came to the Park in 1882 as Assistant Superintendent, built the Cottage Hotel at Mammoth Springs in 1883.

Hofer, T. E., President of the T. E. Hofer Boat Company at Yellowstone Lake since 1907. "Billy" Hofer has been connected with the Park as guide since the early 80's.

Holm, "Tex," Cody, Wyoming, conducted camping parties through the Park several seasons. In 1912 he began to operate in connection with the hotels.

Howard, Gen. O. O., in command of government troops pursued Chief Joseph and the Nez Percé Indians on their hostile raid through Yellowstone Park in 1877.

Hoyt, Hon. John W., Governor of Wyoming, visited the Park in 1881 with the view of opening a road from the southeast. Kepler Cascade near the Upper Basin was named for his son.

Huntley, S. S., succeeded Mr. Geo. W. Wakefield in the stage business in 1892 and was president of the Yellowstone National Park Transportation Company from then until his death in 1901.

Jones, Capt. W. A., commanded an expedition through the Park in 1873, which made its exit through the Absaroka Range east of Yellowstone Lake, at a point now called Jones Pass. In the early 90's he had charge of the road construction as U. S. Engineer.

Joseph, Chief, headed the Nez Percé Indians in their raid through the Park in 1877 entering at the Western, or Madison River Entrance, and making exit at the northeast corner of the Park.

Klamer, H. E., has been identified with the Park since the early 80's. He was proprietor of a general merchandise and curio store at the Upper Geyser Basin until his death in 1914.

Langford, N. P., member of the Washburn-Langford Expedition of 1870; he was the first Superintendent of the Park serving from May 10, 1872 to April 18, 1877. It was largely through his efforts that the region was made a National Park.

Ludlow, Capt. Wm. M., made a report on the Park in 1875 and procured accurate measurements of the Upper and Lower Falls of the Yellowstone.

Lyall and Henderson were identified with the Park since the early 80's. They conducted until 1914, the Postoffice and general store at Mammoth Hot Springs.

Meldrum, Judge, John W., has been U. S. Commissioner in the Park since 1894 with headquarters at Mammoth Hot Springs.

Miles, A. W., president of the Wylie Permanent Camping Company, has been identified with the Park for many years.

Moran, Thos., artist with Dr. Hayden's Expedition in 1871. Artist Point at the Canyon was named in his honor.

McCartney, J. C., took advantage of the squatter's right and located a claim embracing the Mammoth Hot Springs in 1871, prior to the Act of Dedication, and built thereon the first building in Yellowstone Park. The government later reimbursed him to the extent of $5,000 for his rights and improvements.

Nichols, W. M., is president of the Yellowstone Park Boat Company and is connected with the Yellowstone Park Transportation Company.

Norris, Col. P. W., second Superintendent of the Park, served from April 18, 1877, to Feb. 2, 1882; he explored, described and opened up to tourists Norris Geyser Basin and constructed the first wagon roads throughout the Park to the various objects of interest.

Powell, J. D., president of the Shaw & Powell Camping Company, has been identified with the tourist business in Yellowstone Park for many years.

Pryor, Mrs. A. K., under the firm name of Pryor & Trishman, operates the Park Curio Shop at Mammoth Hot Springs.

Raynolds, Capt. W. C., of the Corps of Topographical Engineers, U. S. Army headed a government expedition into the region in 1859 for the purpose of determining the sources of the headwaters of the Missouri River.

Reamer, Robt C., the architect who designed and constructed the unique Old Faithful Inn at the Upper Basin in 1903; also the Colonial Hotel at the Lake, the Depot at Gardiner, and in 1911 the new Grand Canyon Hotel.

Roosevelt, Pres. Theodore, assisted in laying the cornerstone of the Arch at Gardiner in April, 1903.

Ryker, J. N., the first observer in the Park in charge of the Weather Bureau Station established at Mammoth Springs in 1903.

Schwatka, Lieut. Fredrick, of Arctic fame, organized the first winter expedition into the region in January, 1887, but was unable to proceed further than Norris. Mr. F. Jay Haynes and Scout Wilson with two assistants made the entire trip around the Park.

Shaw, Chester, is a member of the Shaw & Powell Camping Company, which conducts a transportation line and a system of permanent camps.

Shaw, Jessie E., is secretary of the Shaw & Powell Comping Company with headquarters in Livingston, Montana.

Shaw, Walter, son of Capt. Shaw, who for many years conducted the steamer Zillah on Yellowstone Lake. He is a member of the Shaw & Powell Camping Company.

Sheridan, Gen. Phil., arranged for the escort of the Washburn-Langford Expedition; headed an expedition through the Park in 1882 with Mr. John McCollough as his guest; in 1883 he accompanied President Arthur on his trip. A mountain near Yellowstone Lake bears his name.

Stevenson, James, was Dr. Hayden's right hand man. His name is perpetuated by an island in Yellowstone Lake and a mountain in the adjoining range.

Stuart, James, from Montana headed a large prospecting party near the Park in 1864.

Vest, Senator Geo. G., drew up a bill, which was passed by Congress in 1884, to protect the Park, authorizing the Hon. Secretary of the Interior to grant restricted privileges in the Park for business purposes. He was a member of President Arthur's party in 1883.

Wakefield, Geo. W., Wakefield & Hoffman of Bozeman, Mont., established in the early 80's the first stage line in the Yellowstone Park. They were succeeded by the Yellowstone National Park Transportation Company in 1892.

Wald, Andy, sand artist, originated in 1888 the idea of filling bottles, showing pictures of animals, geysers and scenes with the different colored sands of the Park.

Washburn, Gen. H. D., Surveyor-General of Montana, headed the Washburn-Langford Expedition of 1870. Mt. Washburn, the observatory of the Park, was named in his honor.

Waters, E. C., was general manager of the Yellowstone Park Association from 1887 to 1890; and from 1890 to 1907 operated a line of steamboats on Yellowstone Lake.

Whittaker, George, succeeded Lyall and Henderson in 1914, as proprietor of the curio and postoffice at Mammoth Hot Springs.

Wylie, W. W., of Bozeman, Montana, in 1890, established the Wylie Permanent Camps throughout the Park.

Yancey, John, in 1871 established in Pleasant Valley near Baronett's bridge on the trail from Gardiner to Cook City (a mining camp just east of the Park), a stopping place for travelers; and lived there until his death in 1905.

THE YELLOWSTONE PARK AND HOW IT WAS NAMED.

The Devil was sitting in Hades one day,
In a very disconsolate sort of a way.
One could tell from his vigorous switching of tail,
His scratching his horn with the point of his nail,
That something had gone with His Majesty wrong,
The steam was so thick and the sulphur so strong.
He rose from his throne with a gleam in his eye,
And beckoning an agate-eyed imp standing by,
Commanded forthwith to be sent to him there
Old Charon, employed in collecting the fare
Of the wicked, who crossed the waters of Styx,
And found themselves soon in a deuce of a fix.

Old Charon, thus summoned, came soon to his chief.
As the Devil was angry, the confab was brief.
Says the Devil to Charon, "Now, what shall I do?
The world it grows worse and grows wickeder, too;
What with Portland, Chicago, Francisco, New York,
I get in my mortals too fast for my fork;
I haven't the room in these caverns below,
St. Peter, above, is rejecting them so.
So hie you, my Charon, to earth, far away.
Fly over the globe without any delay,
And find me a spot quite secluded and drear,
Where I can drill holes from the center in here.
I must blast out more space; so survey the spot well,
For the project on hand is the enlargement of Hell.

"But recollect one thing, Old Charon, when you
Can locate the district where I can bore through,
There must be conveniences scattered around
To carry on business when I'm above ground.
An 'ink-pot' must always be ready at hand
To write out the names of the parties I strand.
There must be a 'punch bowl,' a 'frying-pan,' too.
A 'cauldron' in which to concoct a 'ragout.'
An 'old faithful' sentinel showing my power
Must shoot a salute on the earth every hour,
And should any mortal by accident view
The spot you have chosen, why, this you must do:
Develop a series of pools, green and blue,
That while these poor earth bugs may beauties admire
They'll forget that below I'm poking the fire.
Now fly away, Charon, be quick as you can,
For my place here's so full that I can't roast a man.

To earth flew fleet Charon, to regions of ice;
He found these too cold—so away in a trice
He sought a location in Africa's sands,
He prospected, and finding too much on his hands
He cut out Australia, Siberia, too,
The north part of China—no! they would not do;
Till just as about to relinquish the chase
He stumbled upon a most singular place.
'Twas deep in the midst of a mountainous range,
Surrounded by valleys secluded and strange,
In a country the greatest, the grandest, the best
To be found upon earth—America's West.
Here the crust seemed quite thin and the purified air,
With the chemicals hidden around everywhere,
Would soon make the lakes that the Devil desired;
So he flew to Chicago and there to him wired:
"I've found you a place never looked at before;
You may heat up the rocks, turn on water, and bore."

Then the Devil with mortals kept plying the fire,
Extracting the water around from the mire,
And boring great holes with a terrible dust,
Till soon quite a number appeared near the crust.
Then he turned on the steam—and lo! upward did fly,
Through rents in the surface, the rocks to the sky.
Then with a rumble there came from each spot,
Huge volumes of water remarkably hot,
That had been there in caverns since Lucifer fell—
Thus immensely enlarging the confines of Hell,
And it happens that now when Old Charon brings in
A remarkable load of original sin,

That His Majesty quietly rakes up the coals,
And up spouts the water, in jets, through the holes,
One may tell by the number of spurts when they come,
How many poor mortals the Devil takes home.

But Yankees can sometimes, without doing evil,
O'ermatch in sagacity even the Devil.
For not long ago Uncle Sam came that way
And said to himself, "Here's the Devil to pay.
Successful I've been in all previous wars;
Now Satan shall bow to the Stripes and the Stars.
This property's mine, and I hold it in fee;
And all of this earth shall its majesty see.
The deer and the elk unmolested shall roam,
The bear and the buffalo each have a home;
The eagle shall spring from her eyrie and soar
O'er crags in the canyons where cataracts roar;
The wild fowls shall circle the pools in their flight,
The geysers shall flash in the moonbeams at night,
Now I christen the country—let all nations hark!
I name it the Yellowstone National Park."

                                    WM. TOD HELMUTH.

Grand Canyon, August, 1894.

DISTANCES.

From Western Entrance: **Miles**
Yellowstone, Mont. (W. B.) to Fountain Hotel (F. H.).. 20
Fountain Hotel to Upper Basin (U. B.)................. 9
Upper Basin to West Thumb of Lake (W. T.).......... 19
West Thumb to Lake Hotel (L. H.).................... 15
Lake Hotel to Canyon Junction (C. J.)................ 16
Canyon Junction to Norris Basin (N. B.).............. 11
Norris Basin to Yellowstone, Mont. (W. B.)........... 27

From Northern Entrance:
Gardiner City (G. C.) to Mammoth Springs (M. S.).... 5
Mammoth Springs to Norris Basin (N. B.).............. 20
Norris Basin to Fountain Hotel (F. H.)................ 20

From Eastern Entrance:
Cody, Wyo. to Pahaska................................ 55
Pahaska to Eastern Entrance.......................... 2
Eastern Entrance to Lake Hotel....................... 28

YELLOWSTONE PARK TRAVEL.

(COMPILED FROM REPORTS OF THE VARIOUS SUPERINTENDENTS)

From 1872 to 1894 no complete records were kept including all visitors. Estimates range from one to five thousand each year.

YEAR	Persons	YEAR	Persons
1895...................	5,438	1906	17,172
1896...................	4,659	1907...................	16,414
1897...................	10,680	1908...................	18,748
1898...................	6,534	1909...................	32,545
1899...................	9,579	1910...................	19,575
1900...................	8,928	1911...................	23,054
1901...................	10,769	1912...................	22,970
1902...................	13,433	1913	24,929
1903...................	13,165	1914...................	20,250
1904...................	13,127	1915...................	51,895
1905......	26,188		

ALTITUDES OF PRINCIPAL MOUNTAINS.

	Feet		Feet
Avalanche Pk.......	10,550	Mt. Langford.........	10,902
Bunsen Peak........	8,775	Quadrant Mt........	10,127
Cathedral Pk........	10,650	Mt. Sheridan........	10,385
Electric Peak........	11,155	Mt. Stevenson.......	10,350
Mt. Chittenden......	10,150	Table Mt...........	10,850
Everts..........	7,600	Top Notch Pk..	10,050
Holmes..........	10,350	Mt. Washburn.... ...	10,388

Geysers, Springs, Terraces, Etc., Described Herein.

PUBLICATIONS

The official photographer has prepared for sale from his original negatives many series of photographs in various sizes and styles of Yellowstone National Park.

Plain, Sepia, Colored and Hand-Painted

---

also   *Post Cards*
         Playing Cards
         Souvenir Spoons
         Games
         Haynes Guide Book
         Postpaid 56c
    "As Necessary to the Traveler as His Ticket"

For Full Information—

Haynes

YELLOWSTONE PARK

YELLOWSTONE PARK, WYOMING - ST. PAUL, MINNESOTA

SERVICE

THE OFFICIAL
PHOTOGRAPHER

SOLICITS ORDERS
FOR

*Developing
and
Printing*

IN HIS
LABORATORIES

---

*Groups and Individuals
Photographed*

In order that the most may be gotten out of your films of geysers, terraces and other thermal phenomena, it is strongly recommended that the developing be done by thoroughly experienced operators in this special class of photographic work.

**Haynes
Picture Shop**
at the
Upper Geyser Basin

Haynes

YELLOWSTONE PARK

**Haynes
Photo Stands**
in the
Canyon and Mammoth
Hotels

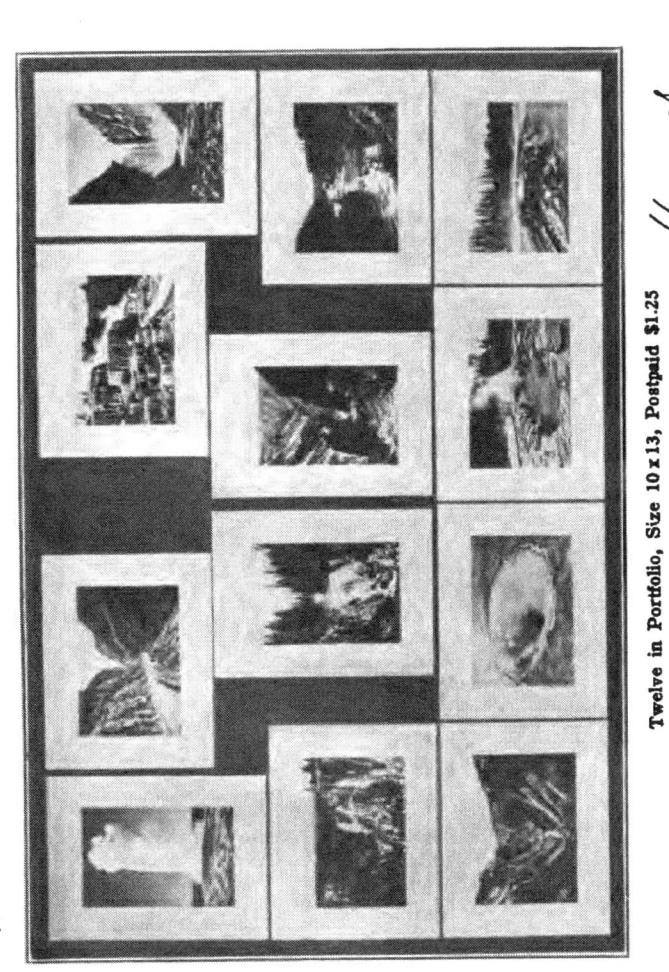

Twelve in Portfolio, Size 10 x 13, Postpaid $1.25

The best souvenir to be had of Yellowstone National Park is this "Photo Color Set" of twelve masterpieces in full color.

Haynes

YELLOWSTONE PARK

CPSIA information can be obtained
at www.ICGtesting.com
Printed in the USA
BVHW031824111022
649159BV00004B/150